The Pandora Series

XXIII

Books in the PANDORA SERIES focus on technology and society – possibilities, risks and uncertainties.

In Greek mythology, Pandora was given a box by the gods but told not to open it. Overcome by curiosity, she opened it anyway. Immediately, all kinds of trials and sufferings flew out over the world. The only thing that Pandora managed to keep in the box was Hope, which is why this has never abandoned humankind.

Editor: Boel Berner

COMMUNICATING CARE

Lisa Lindén

Communicating Care

The Contradictions of HPV Vaccination Campaigns

Arkiv Academic Press

Arkiv Academic Press is an imprint of

Arkiv förlag
Box 1559
SE-221 01 Lund
Sweden

STREET ADDRESS Lilla Gråbrödersgatan 3 c, Lund
PHONE +46 (0) 46 13 39 20

arkiv@arkiv.nu
www.arkiv.nu

A list of Arkiv Academic Press titles can be found in the last pages of this book. For up-to-date information on distribution and available titles, please visit:

www.arkivacademicpress.com

Cover design by Lars Jacobsen
Cover illustration by Charlotte Ewing

First edition by Arkiv förlag 2016
Arkiv Academic Press international edition 2016
For print information, see the back page of this copy
ISBN: 978 91 980854 7 1
ISSN: 1404-000X

Contents

Acknowledgements

Doing a PhD is a collaborative effort. This thesis owes its existence to endless conversations I have had with different people, and to all their encouragement and support.

First of all, I would like to thank all my interviewees at Bredland and Mittland County Councils for agreeing upon meeting with me, and for all their help. I'm truly thankful!

When I first met Ericka Johnson, my main supervisor, I was an undergraduate student at Gothenburg University. My big dream at the time was to do a feminist medical sociology PhD. Ericka was giving a lecture on Viagra discourses, and I was just completely amazed about the fact that someone from my own university was doing exactly the kind of research I wanted to do. So I summoned up enough courage to talk to her during the coffee break. I felt like I was rambling nonsensically, but, and as I since then have come to learn she always does, Ericka responded with enthusiasm and encouragements. A couple of months later we met again when I had become a MA Sociology student and Ericka was teaching a week of medical sociology (alias more Viagra nerdiness). I'm so thankful for that she later on encouraged me to apply for the PhD student position I now am about to finalize. So, thank you Ericka for believing in me. As a PhD student, I have truly appreciated your constant engagement, generousness and support, and your hands-on and constructive comments have always helped and encouraged me to develop my thinking, analysis and writing. For everything: I am deeply grateful!

I have been lucky also to have a great co-supervisor in Claes-Fredrik (CF) Helgesson. From my first day as a PhD student, CF has guided me in the often fun and rewarding, but sometimes so confusing and stressful, life of being a PhD student. CF, for that I am highly thankful. My texts have benefited greatly from your tentative, knowledgeable and thorough readings. I am also grateful that you (and Francis Lee) gave me the opportunity to work as a research assistant in the *Trials of Value* project. Thanks for it all.

I also want to acknowledge the members of my European Research Council (ERC) research group *Prescriptive Prescriptions: Pharmaceuticals and "Healthy" Subjectivities* (PPPHS): Cecilia Åsberg, Ali Hanbury, Ericka Johnson, Oscar Javier Maldonado Casteñada, Tara Mehrabi and Celia Roberts. It has been truly a pleasure to be a part of such an inspiring and welcoming feminist collective of researchers! Thanks for workshops, Skype sessions, collaborations, PhD courses and dinners. Also thanks to the ERC for making this collaboration possible.

Thanks to the PPPHS project I had the opportunity to spend two early summer months at Lancaster University in 2012. In Lancaster I would especially like to thank you Celia Roberts and Vicky Singleton for great inspiration and encouragements, and for taking the time to discuss and read my work. I have also highly appreciated your support in the postdoc application writing process during the last couple of months. Also, I want to thank my fellow HPV vaccination PhD students Ali Hanbury and Oscar Javier Maldonado Casteñada. During those months in Lancaster, and during many times afterwards, we have shared thoughts, ideas and concerns over cups of coffee, beers and walks. It has been great to do this journey together with you. In Lancaster I would also like to thank Mette Kragh-Furbo, Felipe Raglianti and Li-Wen Shih, and the teachers and participants of the Feminist Technoscience Studies summer school. I had the great pleasure of visiting Lancaster at the same time as two other PhD students from Sweden: Kristina Lindström and Åsa Ståhl. It was so great to share the Lancaster experience with you, Kristina and Åsa!

In 2015 I spent five months as a visiting scholar at the Science Studies Program and Catalyst lab at University of California, San Diego (USCD). This visit came to be highly formative for how I wrapped up the thesis. Big thanks to Lisa Cartwright for inviting me, and for being a great host. Lisa, I'm very thankful for constructive feedback, and for all the opportunities you have given me – not at least in relation to the new journal *Catalyst: Feminism, Theory, Technoscience*. Special thanks also to Tania Doles and Cristina Visperas for the friendship, and for all our enjoyable conversations, to Marisa Brandt for recommendations related to communication, media and STS, and to Brian Goldfarb for encouraging me to dig deeper into ethics of care.

Tema Technology and Social Change (Tema T) has been such a great place to do a PhD. The daily "fikas", lunches around the round table, weekly seminars and chitchats on the way to buy a cup of coffee are just a few of all the things that make Tema T into such a great working environment. I would especially like to thank the participants of my two research programs at Tema T: *Technology, Practice, Identity* (P6) and *ValueS: Science, Technology and Valuation Practices*. From P6 I would like to thank Haris Agic, Boel Berner, Elin Björk, Jelmer Brüggeman, Jenny Gleisner, Hannah Grankvist, Lisa Guntram, Sonja Jerak-Zuiderent, Ericka Johnson, Corinna Kruse, Oscar Javier Maldonado Casteñada, Erik Malmqvist, Anna Morvall, Alma Persson, Sarah Jane Toledano, Kristina Trygg and Kristin Zeiler. From ValueS I would like to thank Réka Andersson, Jeffrey Christensen, Ivanche Dimitrevski, Claes-Fredrik Helgesson, Lotta Björklund Larsen, Oscar Javier Maldonado Casteñada, Johan Nilsson, Nimmo Osman Elmi, Karin Thoresson, Bistra Vasileva, Steve Woolgar and Teun Zuiderent-Jerak. There are also a few amongst the former P6 and Values participants I would like to thank: Maria Björkman, Baki Cakici, Maria Eidenskog, Malin Henriksson, Linus Johansson Krafve, Francis Lee and Anna Wallsten. To all these current and former participants: thank you for fun, enriching and friendly conversations about STS, feminist theory and, perhaps above everything, the everyday dos and don'ts of writing academic texts. You all have taught me how to be a researcher!

Many have read my texts over the years. I am thankful for a great reading from my 60% opponent, Isabelle Dussauge, and from the reading group, Teun Zuiderent-Jerak and Anette Wickström. Your well-informed, thorough and encouraging comments helped me believing in my project after a period of doubt, and guided me through all the decisions I had to make. My final seminar/90% opponent, Kerstin Sandell, provided insightful and constructive comments that have helped me immensely during the last months of wrapping up the thesis. Thanks for helping me to kill my darlings, and for encouraging me to further develop what I really care about! I am also grateful for knowledgeable and supportive readings from the final seminar committee Nina Lykke, Harald Rohracher and Anna Sparrman. Also thanks to Maria Eidenskog, Anna Wallsten and Kristin Zeiler for reading parts of my thesis before the final submission.

When I started in the fall of 2011, I was the only new PhD student at Tema T. This, however, has not meant that I have been lonely. Quite the opposite! Early on the D-10 PhD student group adopted me into their group, and for that I am grateful. During the years, our support meetings and countless coffee breaks have helped me during difficult times, and have been a source of everyday joy. Thank you Réka Andersson, Maria

Eidenskog, Linnea Eriksson, Mattias Hellberg, Linus Johansson Krafve, Katharina Reindl, Hanna Sjögren, Josefin Thoresson and Anna Wallsten. Special thanks to Anna and Réka for "pepp" lunches and coffee breaks, and for being experts in giving hands-on advice during moments of crisis. During the last months, your everyday support has done wonder for my wellbeing. Also thanks to Hanna for the friendship; I highly cherish our conversations about feminism, life and academia over ice creams, dinners, beers, and nowadays Facebook and Skype. You are missed in Linköping. Finally, I want to thank Maria. I cannot imagine a better officemate than you, Maria! Our countless walks to the campus cafés for our daily coffee, our conversations about care and STS, all the dinners, and our joint big love for dogs; all of it has meant a lot to me. I'm happy that you have become a close friend of mine.

Also another group of PhD students adopted me from the very start: the Gender Studies unit/Tema Genus D-11 group. Thank you Line Henriksen, Marie-Louise Holm, Desirée Ljungcrantz, Tara Mehrabi, Marietta Radomska and Helga Sadowski. I'm thankful I got the opportunity to take PhD courses with you our first semester. I learned a lot, and I gained a great new collective of fun, sharp and supportive friends and feminists in my life. Thanks for all the parties, reading groups and dinners over the years. Especially thanks to Line and Marietta for being my Norrköping comrades-in-crime! Thank you for bats, ghosts and weird inside jokes, and for countless visits to the pub Broadway. Also thank you Tara for always being such an encouraging and caring friend; you are someone to lean on.

There are some additional Tema T people I would like to acknowledge. Thank you Sonja Jerak-Zuiderent for reading my texts carefully, and for being such a warm, kind and generous person. Our conversations about care and commitments have meant a lot to me, and for this thesis. And thank you Anna Morvall for all the dinners, walks with the dog Dino, and for the friendship. From the very start of my PhD studies you have been someone to lean on, and a very important person in my life. For that I'm profoundly thankful. At Tema Genus I would like to thank Cecilia Åsberg, Nina Lykke and Margrit Shildrick for support and insightful comments on my work, and the Posthumanities hub research group members for inspiring seminars and workshops. Outside of Tema, I would like to thank Doris Lydahl for countless conversations in diverse places such as Copenhagen, Denver and Gothenburg, and for being my first STS friend!

A number of people have helped significantly with the administrative practicalities of academia. Thank you Eva Danielsson, Ian Dickson, Carin Ennergård, Josefin Frilund and Camilla Junström Hammar. Also thanks to Pat Baxter for great proofreading, and to my graphic designer Charlotte Ewing for making the cover look so good. At Arkiv förlag/ Arkiv Academic Press I would like to thank Boel Berner and David Lindberg for excellent editing work, and for turning my manuscript into a book!

As a PhD student I have had the pleasure of teaching at the Society and Culture Analysis (SKA) program together with Anna Bredström. Thank you Anna for giving me the opportunity to teach about things I love to talk about, and for through your enthusiasm, sharpness and friendship making the experience even more pleasurable!

This thesis also owes its existence to all my loved friends outside of academia. Thank you Anna B, Anna N, Cissi, Daniel, Johanna, Josefin, Karin, Liv and Majja for the fact that you have kept reminding me that there is a life also outside of academia, and for making that life so enjoyable. Thank you for all the care and support, for lifting me up when I have been down, and for all the fun times.

My family has always supported my passion for studies and research. Thanks to my mom Åsa and Roger for their support and love, and for always, always believing in me. It means the world, and more. Agnes, my sister – thanks for everything. I am lucky to have you as one of my best friends. Thanks for all the dinners in our flat, walks along Strömmen, "jympa" classes and Meze restaurant visits. Erik, my brother – thanks for conversations over beers and dinners in Gothenburg, and for being a great brother. Also thanks to my granddad Caj for support, and for being the coolest granddad in the universe! Special thanks also to the dog Kate for relaxing walks in the forest when I have needed it the most. Finally, I would like to acknowledge my dad Thomas and my grandma Gunnel, who are not with us any longer. I will always be deeply grateful and happy for everything you have given me.

Norrköping, April 2016
Lisa Lindén

1. INTRODUCTION: HPV Vaccination Campaigns and Temporalities of Care

"Have you gotten vaccinated?" a school nurse asks two teenage girls in a video. One of the girls answers yes, the other one says no. As a response to this, the school nurse turns her attention from the girls, looks into the camera and says to the audience "Have *you* thought about doing it? It gives really good protection against cervical cancer". In turn, on a Facebook site, I am asked, "Who do you care about?" This is followed by an encouragement to share a message about human papillomavirus (HPV) vaccination to others I care about. Facebook encourages me to share the message "Get vaccinated against cervical cancer now!" Finally, when I am standing waiting for the bus, I encounter a poster of a young man named Lukas (Figure 8, page 178). He looks steadily into the camera with a serious facial expression. He tells the viewer – me – about his experiences of having a mother with cervical cancer. "The only thing I wanted was that I could be sick instead of her", he says. The bus arrives, and I still have Lukas's words ringing in my head.

These examples are linked to three different HPV vaccination campaigns in Sweden: an "HPV app" campaign and two different campaigns both entitled "I love me". In the campaigns, different forms of care are figuring. In the app, the video of the school nurse promotes getting vaccinated as a matter of teenage girls caring for themselves, and the school nurse as someone caring for girls. The Facebook campaign site, in turn, informs me that sharing the "get vaccinated now!" message on Facebook is an act of care; sharing is caring. Moreover, through the cancer narratives, I encounter people (such as Lukas) emphasizing their care for their cervical cancer afflicted relatives.

In the three campaigns, care is presented as something temporal. The HPV app and the Facebook site encourage people to get vaccinated *now* to prevent *future* cervical cancer, and in the cancer narratives relatives tell stories about memories of pain and grief, and about a fear of their relatives in the future getting cancer back. My study zooms in on these three HPV vaccination campaigns to ask questions about care, and especially about care as a temporal matter.

In Sweden, and in the majority of countries with national vaccination programs, HPV vaccination is offered free of charge to teenage girls, to prevent cervical cancer and to prevent genital warts. The two HPV vaccines currently on the market, *Cervarix* and *Gardasil*,[1] are through campaigns promoted worldwide as vaccines against cervical cancer (Wailoo et al. 2010). In many of these campaigns, girls are encouraged to get vaccinated as an act of caring for themselves (Polzer and Knabe 2009; Davies and Burns 2014), and parents (along gendered lines, often mothers) are encouraged to care for their daughters through vaccination (Connell and Hunt 2010).

Also, outside of the context of HPV vaccination, campaigns are extensively used by public organizations to encourage people to adopt specific medical treatments (such as vaccinations), and/or to start to live healthier lives. They encourage a "care of the self" (Serlin 2010b: xxi). Moreover, and as exemplified by the empirical examples brought up above, campaigns often articulate a "care for others" along gendered lines. Also, they tend to depict care as a matter of acting now to safeguard a healthy future (Cartwright 2013; Coleman 2015). Thus, in encouraging people to change their health behaviors or to help changing others, campaigns may include moralizing assumptions emphasizing that citizens *should* act *now* in accordance with the communicated message, and in line with societal expectations (see e.g. Lupton 1995; Moulding 2007; Crawshaw 2012). However, at the same time, health campaigns can also include and enable less moralizing and normalizing notions about what care might be, notions that allow for multiplicity, contingency and uncertainty (Fraser and Seear 2011). Thus, health campaigns are not, and do not have to be, only *one* thing.

I situate this study within the interdisciplinary field of science and technology studies (STS), and, more specifically, within feminist STS studies on care (see e.g. Puig de la Bellacasa 2010, 2011; Martin et al. 2015). STS as a field is interdisciplinary and diverse, but it can broadly be

1. Cervical cancer is associated with specific HPV types, most frequently types 16 and 18. These are the two types *Gardasil* and *Cervarix* vaccinate against. In addition to this, Gardasil vaccinates against HPV types 5 and 11, which are strongly associated with the development of genital warts (National Cancer Institute 2016).

explained as focused on how culture, politics and society form science and technology, and, conversely, how science and technology form culture, politics and society. The field takes an interest in the social and material complexities of scientific and technological discourse and practice. Using an STS approach makes it possible for me to attend to the contingencies, contradictions and materialities of HPV vaccination campaigns.

Drawing upon feminist STS studies on care, I show how HPV vaccination campaigns include, and enable, different *matters of care* (Puig de la Bellacasa 2011). Care is approached as a relational *doing*; care is made through relations, and in relations. Thinking about care in this way allows me to attend to how care is made in a multitude of different ways in, and through, the studied campaigns and their practices. This allows me to keep open toward what care in this context might be and become, rather than restrict my analysis to practices of self-care and care for others. The feminist STS approach to care also makes it possible for me to pay close attention to normative and exclusionary ways of doing care in the campaigns *and* to those that open up for more caring and livable practices.

The matters of care articulated *in* the campaign material are not the only ones I attend to. For example, on the "I love me" Facebook campaign site already mentioned, lay people – or as I will refer to them, *publics* – participate in discussing HPV vaccination and the "I love me" campaign, and in doing care. Instead of simply confirming to the message of "get vaccinated against cervical cancer now!", publics on the Facebook site encouraged others to take care of their lives by taking the time to think it over before making a vaccination decision. That is, *through* the digital practices of this campaign, *temporalities of care* other than the ones visible in, for example, campaign images, are also made possible.

Moreover, the participation of actors on the Facebook site does not only include publics. Communication on this site is enabled and mediated through a range of material devices, such as Facebook social buttons (for example, the like and share buttons). By drawing upon STS insights concerning the importance of including material objects as participants in doings of care (e.g. Mol et al. 2010a), I attend to how such devices take part in doing matters of care. Moreover, through interviews with county council professionals that have worked with the campaigns, I discuss matters of care brought up and reflected upon in conversations about the campaigns. Thus, STS approaches to care make it possible for me to examine closely human *and* nonhuman actors as participants in the making of different matters of care.

Finally, I attend to the care made possible through *my relations* with the actors and worlds I study. In doing so, I try to take seriously that what I focus on in this study has implications for how care is being made. I do not want to make absent that I, as the researcher, care for certain things. Following this, I do find it problematic that health campaigns are often done as moralizing endeavors: you *should* get vaccinated, you *should* care! I also find it worrying that HPV vaccination campaigns often mobilize gendered assumptions to encourage girls to make up their minds. I care strongly for these issues; they trouble and worry me. These things I attend to in this study.

However, by also slightly shifting focus, I argue that a plentitude of other things is already part of the story. In learning from other researchers working on care (see e.g. Mol 2008; Mol et al. 2010a; Puig de la Bellacasa 2011, 2015), by caring about these other things – to flesh them out, to strengthen them – I believe it is possible to foster matters of care that hopefully can enable more caring health communication practices. By holding on to moments in the empirical material which open up space and time for alternative, and more inclusive and caring, matters of care, I work with an approach where I try to "slow down" and disrupt calls for a need of getting vaccinated *now*. Formulations of care as an urgency ("get vaccinated against cervical cancer now!") often close down possibilities for alternative action, and therefore it is important to try to foster and strengthen other matters of care.

Concretely I do this by attending to neglected, marginal, absent and alternative matters of care. By allowing them a center space, through this study I aim to tell complicating stories about diverse, and sometimes contradictory and conflicting, matters of care. Some of the matters of care I emphasize might seem trivial in comparison to the "bigger issues" involved. Learning from, for example, Maria Puig de la Bellacasa (2011, 2014), this is precisely the reason why seemingly trivial matters need attention. This mode of attention, I hope, can help disrupt and unsettle some of the normative and exclusionary ways in which care is being done. Using this approach, I try to practice care in a responsible manner that helps enabling more caring relations.

I attend to *several dimensions of care*: the care articulated in, and through, different campaign media, the care enacted in campaign practices, and the care I take part in articulating and fostering. Following this, I aim to hold on to care as something simultaneously promising, risky and, troubling. I attend to *the promises* and *troubles* of care. This means that I do not believe care per se is something good or desirable. As others have shown, care can be about social control and governance (Davies and Horst 2015), gendered relations of power (Murphy 2015; Viseu 2015), and a moralization of people's behavior; if only *you* would care! (Puig de la Bellacasa 2012). At the same time, and as already indicated, I also emphasize that attending to care can help foster alternative ways of thinking about (doing) ethics and politics, or as by following Puig de la Bellacasa (2011), I will discuss it as *ethico-politics*. This, at least partly, has to do with the fact that care comes with connotations of commitment, affectivity and interdependence. By "thinking with care" (Puig de la Bellacasa 2012), care, with all its possible potential, trouble and riskiness, is also made present and allowed space. Doing so, I emphasize, makes possible discussions around the multilayered (inclusionary and exclusionary) *politics* of care.

Aim and research questions

The aim of this study is two-fold. First, the aim is to show how matters of care are *articulated* and *mediated*, in, and through, HPV vaccination campaigns, and by professionals working with the campaigns. Secondly, the aim is to show how attending to matters of care as an ethico-political mode of attention in a context of HPV vaccination campaigns can trouble normative and exclusionary matters of care. To be able to do so, the study pays attention to predominant articulations and mediations, as well as to absent, marginal, neglected and alternative ones. Focus is especially put on how alternative temporalities of care may disrupt normative and exclusionary links between care and time. In approaching health campaigns through this theoretical approach, the study aims to theoretically and conceptually contribute to STS discussions on matters of care. This leads to the following research questions:

1) How, and what, matters of care are articulated and mediated in the campaigns?
2) How, and what, matters of care are articulated by county council professionals working with the campaigns?
3) By attending to absent, marginal, neglected and alternative articulations and mediations, what other matters of care are made present?
4) By attending to different temporalities of care, how is it possible to trouble and disrupt normative and exclusionary links between care and time?

5) How can these findings contribute to STS discussions on matters of care in technoscience?

To answer these questions, and partly as already mentioned, I use a combination of methods. As is explained further in Chapter 3, I combine a close reading method with an STS device perspective. Through this, I analyze how care is visually and textually presented in the campaign material and how this is enabled by, and articulated through, different digital and non-digital material devices.

The next question relates to the aim of showing how professionals discuss working with the campaigns, and is based on interviews with them. Through this question, the study explores how professionals' articulations can be interpreted as involving matters of care.

By concentrating on marginal, absent, neglected and/or alternative articulations in both the campaign material and the interviews, I focus on the matters of care enabled and staged through such mode of attention. Additionally, in focusing on temporalities of care in the campaign material and in the interviews, I discuss the politics of temporalities of care present in my material. Finally, through the last question, the study's theoretical contribution to STS, and especially to feminist STS, is discussed.

Cervical cancer, Pap smears and HPV vaccines: debates and issues

Before I continue, some context about cervical cancer prevention and HPV vaccination is needed. Since the 1960s all Swedish women between 23 and 60 years old are offered regularly Pap smears (gynecological screening) for the prevention of cervical cancer. Before the introduction of the national screening program approximately 900 women yearly got inflicted by cervical cancer. Since the introduction of the national screening program cervical cancer has gone from being the third largest cancer amongst Swedish women to become the 17th (Bäcklund 2015). The decrease in cervical cancer prevalence in Sweden is equivalent to the situation in other countries with national screening programs (ibid.). Screening programs and HPV vaccines serve as reasons for why in the Global North nowadays cervical cancer is understood as a highly preventable disease (Löwy 2011). At the same time, cervical cancer has become a disease of the Global South where access to screening programs is more limited (Maldonado Castañeda 2015).

Whereas proponents have announced HPV vaccines as the first cures against cancer (Wailoo et al. 2010), critical observers have commented on the vaccines as costly and uncertain technologies that might medicalize girls and young women unnecessarily. It has been argued that women perhaps would benefit more from improved Pap-based screening programs than from HPV vaccination (Paul 2016: 194). Researchers have also stressed the problems with, in campaigns and elsewhere, making absent uncertainties about the vaccines' long-term efficiency, and about their possible risks (e.g. possible side effects) (Maldonado Castañeda 2015). Others have stressed problems with reduction of specificities concerning the quite complex links between sexual activities, HPV infections and cervical cancer, including the fact that HPV vaccines are estimated to only protect against 70 percent of all cervical cancer occurrences (Braun and Phoun 2010). It is thus problematic to present HPV vaccines as *cures* against cervical cancer.

A focus on HPV vaccines as cures against cervical cancer also allows for an enactment of HPV vaccines as "girls' vaccines" (Mishra and Graham 2012), something that makes absent boys as possible vaccine recipients (Lindén 2013b). A few countries include boys in vaccination programs to enable prevention of anus, throat and penis cancers in males. In the context of Austria, for example, a discourse of gender neutrality has been drawn upon to encourage parents to vaccinate both their boys and their girls (Lindén and Busse forthcoming).[2]

Not only gendered politics matter in a context of cervical cancer and HPV vaccination. Cervical cancer has a long history of being a disease associated with marginalized groups. Especially assumptions about links between women from marginalized groups, cervical cancer and "sexual promiscuity" have articulated exclusionary cervical cancer discourses and practices (Löwy 2011). Today in Sweden, the national cervical cancer screening program is related to issues of class and ethnicity, this since participation is less frequent amongst foreign-born citizens and amongst the working class (Cancerfonden 2016). As I have discussed in previous work (Lindén 2013a), this situation has made some argue for the need of HPV vaccines, while others have argued for a need to improve the existing screening program. Also concerns have been raised about the fact

2. In March 2016, it was announced in Swedish media that the Swedish Public Health Agency will examine whether it would be possible in the future to include boys in the national vaccination program in Sweden (*Läkemedelsvärlden* 2016). If an inclusion of boys is agreed upon in Sweden, gendered HPV vaccination discourses will likely change.

that exaggerated expectations on HPV vaccines might make less people take the Pap smear (Rehnqvist et al. 2008).

Why study HPV vaccination campaigns?

As the above overview illustrates, HPV vaccination is a phenomenon that includes scientific uncertainties and social and political complexities. Therefore, there is not *a given* "yes" or "no" answer to whether HPV vaccination is needed in Sweden or not. Instead, I argue that what the complexities at hand do make clear is that there is a need for taking the politics, uncertainties and specificities of HPV vaccination practices seriously, including their multilayered problems, risks and/or possibilities.

A focus on HPV vaccination campaigns allows me to attend closely to such uncertainties, specificities and politics. Making HPV vaccination campaigns is anything but straightforward. My interviews indicate that this work includes a wide range of actors and ethico-political issues, debates and tensions that professionals need to handle, navigate and respond to. In putting these matters up front, I emphasize the possibilities and challenges of care in health campaigns; its diverse promises and troubles. An important rationale with my study is therefore to provide input on an ongoing discussion in, and beyond, academia, about the stakes of HPV vaccination campaigns, and of other health campaigns. I want to widen the scope of the discussion to encompass critique that takes into account the situated tensions, considerations and navigations that enable health campaigns to include specific matters of care, and not other. Ultimately, taking the tensions, uncertainties and specificities of campaigns and their practices seriously makes possible for a more situated, comprehensive and fruitful academic and public debate.

A Swedish health care and public health context

Before I further introduce the campaigns concerned in this study, there are a few things about Swedish health care that need to be explained. In Sweden, health care is for the most part funded by taxes,[3] and is connected to national regulations emphasizing citizens' rights to equal

3. There is a maximum amount Swedish citizens need to pay for health care provision, and this is regulated through a system of subsidization. For health care that does not require hospitalization, the maximum sum is 1,100 Swedish Kronor yearly, and for pharmaceuticals it is 2,200 Swedish Kronor yearly (Swedish Dental and Pharmaceutical Benefits Agency 2016).

health care (Swedish Government 1982). More specifically, in the Swedish Health Care Act, it is stated that "the aim of the health care services is good health and equal care for the whole population", and that "those who are in most need of care shall be given priority to access care" (ibid., my translation from Swedish). This is based on three "ethical principles" on which the Swedish Government has decided. *The human dignity principle* postulates "all humans have the right to equal worth and the same rights independent of personal capacities and functions in society" (Swedish Government 1996 (Prop. 1996/1997:60), my translation from Swedish). In turn, *the need and solidarity principle* states "resources should be distributed based on need" (ibid.). Finally, *the cost-effectiveness principle* emphasizes that "in case of a need to choose between activities or intervention, a reasonable relation between cost and effect, based on improved health and quality of life, should be aimed for" (ibid.). It is, however, emphasized that the cost-effectiveness principle should not overrule the others (ibid.). The system is a welfare system, articulating a combination of equality, solidarity and financial matters as ways of realizing good health care provision.

How these principles are implemented differs in different regions of Sweden, where regional county councils and local municipalities organize health care. The county councils are partly independent of the state, and have a mandate to implement national regulations through regional and local adjustments. Both the county councils and the municipalities are responsible for implementing health care services according to the law, but they have different areas of responsibility.

In general the municipalities have responsibility for child vaccinations which are provided free of charge through the school health system. This is also how the general HPV vaccination program provided for girls between 11 and 12 years old is delivered.[4] However, in the matter of vaccination this study is concerned with (the "catch-up" HPV vaccination) responsibility was solely vested in the county councils (more details follow below). At the same time, in their recommendations for the implementation of the HPV vaccination program, the responsible governmental organization, the Swedish Association of Local Authorities and Regions (SKL), encouraged collaboration between the county councils and the municipalities (SKL 2010b). It was emphasized that a catch-up vaccina-

4. The age (11 to 12 years old) of the "target group" is set based on medical findings emphasizing that it is most effective to get vaccinated before "sexual maturity", as HPV is sexually transmitted (Hildesheim and Herrero 2007).

tion through the school health system would provide "considerably better vaccination coverage", and that it was therefore desirable to provide this vaccination also in schools (SKL 2010b).

In Sweden, recent market reforms have transformed the health care system. There are several reforms that could be discussed, but of direct relevance here is the "care choice system" reform that since 2008 has been opted for in Sweden. The Swedish Government has decided that all county councils must organize their primary health care through this system. Even though this reform comes with many specificities, what is needed to know for this study is that the care choice system means that citizens can choose what care provider they want to go to, be it public or private, and that regional governments in Sweden need to implement systems that facilitate such patient choice (Swedish Government 2008 (Prop. 2008/09:74)). As Linus Johansson Krafve (2015: 7–11) explains in his PhD thesis *Valuation in Welfare Markets*, the care choice reform was decided for on the basis of an idea that competition between care providers would improve the level of care, that it would "empower" patients, and that it would enable "cost-effective" health care solutions. Another vital aspect was the idea that this system would generate better care accessibility, as the reform allows for an increased level of care providers on "the health care market" (Swedish Government 2008 (Prop. 2008/09:74)).

A focus on patient choice and empowerment can also be found in current Swedish public health policies, and this further helps to contextualize my case.

> Public health work should first and foremost aim to promote health. The work needs to be formulated on the basis of people's need for integrity and freedom of choice. To promote health is a process that enables people to increase control over their health and to improve their health [...] The government wants to promote the individual's interest, responsibility and capability for taking care of her/his own health. (Swedish Government 2007 (Prop. 2007/08:110: 9).)

A central part of current public health policy in Sweden is, as the quote shows, an emphasis on the individual's own capacity and ambition to promote population health. Here, the state's (and regional government's) responsibility for citizens' health is downplayed in favor of a focus on citizens' responsibility to take control over their health, and to care for themselves.

What can be taken from this is that the health care policies and regulations in Sweden need to be understood as a mixture of welfare and market values. They emphasize needs, solidarity, competition, state responsibility, patient empowerment and free choice, at the same time. The Swedish health care system is, as Johansson Krafve (2015) explains, a welfare *market*.

Bredland and Mittland County Councils and the HPV vaccination campaigns

The three HPV vaccination campaigns this study is concerned with are located in two different county councils, here called Mittland County Council (the "HPV app" campaign) and Bredland County Council (the two "I love me" campaigns). The three campaigns concerned only the catch-up HPV vaccination. This vaccination was decided on since the financial budget allowed for it. Moreover, since previous evaluations from the Swedish National Board of Health stated that a catch-up vaccination would be effective such extended vaccination was recommended (SKL 2010a, 2010b).

A legal conflict over the national procurement process partly contributed to the decision that the national budget allowed for a national procurement of Gardasil. This conflict included the Swedish county councils, governmental agencies and pharmaceutical companies, and resulted in a reduction of the vaccine price. In 2010, it had been decided nationally to purchase the HPV vaccine Cervarix, on the basis that this vaccine was less expensive than Gardasil. It was also stated that procuring the less expensive vaccine Cervarix would allow financially for a catch-up vaccination (SKL 2010a). Despite the fact that Cervarix does not include protection against genital warts, it was stated that the significant price difference between the vaccines could not motivate the national procurement of Gardasil. The pharmaceutical company behind Gardasil (Sanofi-Pasteur MSD), however, lodged an appeal against this decision, as they believed that the contractual period for the procurement was too long to be in accordance with the relevant contractual regulation (Knutson and Öster 2013).

This appeal generated a new national procurement in September 2011, in which SKL decided to procure Gardasil instead of Cervarix. Of importance was that Sanofi-Pasteur MSD had decided to reduce the price of Gardasil for the national procurement. On the basis of this reduction in price, it was stated that the positive health effects of procuring a vaccine that also included protection against genital warts motivated the procurement of the more expensive vaccine (SKL 2010b).

However, on the basis that interpretations of the scientific data used to evaluate the efficiency of the vaccines were seen as questionable, this led to a request for a new appeal, this time from the pharmaceutical company behind Cervarix, GlaxoSmithKline (GSK). The new appeal was, however, not approved by the Swedish courts since it was agreed that the decision to procure Gardasil had been correctly made (Knutson and Öster 2013). Still, this conflict resulted in a reduction of the vaccine price and national procurement of Gardasil instead of Cervarix for both the general vaccination and the catch-up one.

This conflict over the procurement process was extensively covered, and criticized, in the Swedish media. It was stated, for example, that delaying the start of the vaccination could "cost girls their lives" (*Expressen* 2011) and that it is a "scandal that young girls have to wait for the vaccination" (RFSU 2011). Especially criticized was how pharmaceutical companies' commercial and financial self-interests had delayed the process (Lindén 2013a).

As I have already mentioned, the campaigns focused on in this study only concern the catch-up vaccination. This means that they address girls and young women aged 13 to 20 in Mittland County Council, and, since 2012, aged 13 to 26 in Bredland County Council. That Bredland County Council decided to increase the age limit from age 20 to 26 was partly due to the fact that they had a budget that allowed for it. However, it was also related to epidemiological findings which argued for its efficiency in vaccinating up to this age limit (see e.g. Harper and Paavonen 2008). The HPV catch-up vaccination started in early 2012 in all Swedish regions, including in Bredland and Mittland. In Bredland and Mittland (and, to the best of my knowledge, in other regions as well), it will end in 2016.

The catch-up HPV vaccination coverage in both regions is close to the national average, although one of them is a little below, and one is a little above (PHAS 2014). On a national level, no county councils are clearly below the average, but a few are significantly above. Even though the reasons for this most likely vary, one example that was often brought up by my interviewees was that one county council had decided also to vaccinate the catch-up group in schools. This school vaccination was conducted by recruiting retired nurses as volunteer vaccinators. This county council is one of the ones with the highest vaccination coverage (ibid.).

To enable increased care accessibility, Bredland and Mittland County Councils – in contrast to many of the other county councils in Sweden – use a care choice system for the catch-up vaccination. However, including the catch-up vaccination in the care choice system is not the only way to do this, as the catch-up vaccination can also be organized through a system similar to the regular vaccination program. However, and in line with the vision of the national care choice reform, in Bredland and Mittland it was believed that enabling different (public and private) care providers to offer the catch-up HPV vaccination would increase care accessibility, and therefore also vaccination coverage. To further increase vaccination coverage, and in line with the recommendation from SKL (2010b), Bredland and Mittland County Councils encouraged high schools to become authorized as vaccinators, but few did so. Worth noting is that in 2015, Mittland County Council introduced a new strategy for the catch-up vaccination where a collaboration between the county council and high schools enabled vaccination in high schools. This makes the catch-up vaccination in this region similar to the general vaccination scheme. However, as this vaccination was decided after I had conducted empirical research it is outside the scope of this study.

Campaign materialities

Since the catch-up vaccination is part of the care choice system in Bredland and Mittland, girls and young women have to actively find a vaccinator to get vaccinated. This is different both from the general vaccination program, where girls and young women are vaccinated through the school health system, and where the catch-up vaccination is organized in a similar way. When girls and young women need to find a vaccinator, vaccinators need to make sure that they are possible to find. Because of this, Mittland and Bredland County Councils have as part of their assignment to ensure that connection between girls and vaccinators is possible. Since it is not a given for girls and young women to know where to go (for example, in regions where the catch-up vaccination is set up as a school vaccination, information is simply given at school), the county councils have to inform the girls and young women about it. Therefore, they created the campaigns examined in this study.

In their HPV vaccination campaigns, Bredland and Mittland County Councils have worked with what my interviewees sometimes referred to as "non-traditional" participatory health campaign media: digital media (the HPV app and the Facebook "I love me" site), and an "I love me" vaccination trailer that Bredland County Council trawled around high schools in the region to enable school-located vaccination. The campaigns also include several more "traditional" media: posters, pamphlets, movies and "regular" web pages.[5] In this study, focus will be on a few of these: the app, the Facebook site, posters, and textual cancer narratives represented on a campaign web page. The use of digital media for communicative purposes is something that is stated as desirable in the county council communication guidelines. In these, it is articulated that public dialogue as well as "target group" specific communication is important, and that digital (especially social) media can enable this.

The use of digital media for health campaign purposes needs to be contextualized, as such possibilities are currently extensively discussed in health promotion and preventive medicine literature. As in Bredland's and Mittland's guidelines, in this literature, apps and designated Facebook pages related to specific public health campaigns or health behaviors are proposed as a new promising arena for "target group" adjusted communication (e.g. Lefebvre 2009). It is especially emphasized that using digital media has the potential to foster patient empowerment (e.g. Korda and Itani 2013) and public engagement (e.g. Neiger et al. 2012). As an important backdrop for my study, children and teenagers are often referred to as "target groups" that can productively be reached through digital media (Evans 2008; Evers et al. 2013), as digital media are envisioned to "reach youth on their own terms" (Ralph et al. 2011: 48).

Vaccinations are brought up as another area that can productively be promoted through digital media, and especially through social media (Betsch et al. 2012; Wilson and Keelan 2013). While researchers raise concerns about vaccination critics "hijacking", for example, Facebook vaccination campaign sites, they also stress the potential participatory and empowering capacities of social media. It is especially emphasized as an arena that, by enabling expert-citizen dialogue, can counteract, what, in the context of vaccinations, is often articulated as misunderstandings about science. Directly in the context of HPV vaccination, it has been stressed that "using social media tools (e.g. Facebook, Twitter) is [a] key strategy to disseminate accurate information and dispel some of the misinformation that is spread by the anti-vaccine movement" (Zimet et al. 2013: 416). Thus, it is by health communicators envisioned that public

5. As Anders Ekström with colleagues emphasize in the anthology *History of Participatory Media* (Ekström et al. 2011), participatory media has a long history. It is more complex than that participatory media is something "non-traditional", and that other forms of media are "traditional". See especially the anthology's chapter by Ylva Habel (2011) on a participatory public health campaign.

engagement and user interaction will enable effective vaccination communication. Several of the aspects (e.g. public engagement and empowerment) brought up in this literature on using digital media for health communication purposes are themes that will be discussed in my study.

Previous research

My study relates to several different fields and discussions. Being a study on care in a context of HPV vaccination campaigns, it is connected to social science and humanities research on public health in general, and vaccinations and (public) health campaigns in particular. Researchers working in these areas are from different theoretical fields, including, for example, visual culture, media studies, history, sociology and STS. Despite this diversity, since I aim first and foremost to contribute to the field of STS, and primarily to feminist STS, I will particularly focus on such research. Since only a few STS studies exist on health campaigns, when talking about this research, I will also bring up research from other fields. I will discuss, and relate to, research that will help the reader understand my approach and argument. This includes both empirical and theoretical insights from other studies. The majority of matters raised will in one way or another be returned to later in the empirical chapters.

Importantly, as it is a study focusing on a Swedish case, I attend particularly to work from Sweden, and other Nordic countries. Furthermore, since my analysis concerns campaigns related to sexual matters, sexually transmitted infections and female cancer, extra attention will be on these matters. The research overview will end with a section where I relate my study to this previous work.

Public health in STS

Many researchers working within the field of STS have taken an interest in public health, both in Sweden and elsewhere. From different STS perspectives, areas such as HIV/AIDS (Epstein 1996), hepatitis C (Fraser and Seear 2011; Cartwright 2013), nicotine replacement (Elam and Gunnarsson 2012) and, as in my study, cervical cancer prevention (Singleton and Michael 1993; Singleton 1998; Casper and Clarke 1998; Wailoo et al. 2010), have been studied. Importantly, studies show, for example, how public health intersects with power differentials such as gender, race and/or class (Singleton 1995; Epstein 1996), and how public health has become increasingly influenced by the politics of individualized responsibility, lifestyle and risk (Boero 2010). Some researcher also point toward the temporal dimensions of public health, such as how preventive interventions may include a logic of futurity (anticipatory, future-oriented time emphasizing immediacy) which tends to assume that the future always is better, and which promises happiness and health if people act *now* to prevent disease or illness (Adams et al. 2009; Roberts 2015).

Notably, STS scholars have discussed vaccinations as meetings and tensions between medical experts and lay publics (Collins and Pinch 2005: chap. 8; Leach and Fairhead 2007; Bragesjö and Hallberg 2009). Some use this focus to indicate transformed relations between experts and publics, where discourses of parental choice, empowerment and decreased trust in science and public authorities change the current vaccination landscape. For example, in *Vaccine Anxieties*, Melissa Leach and James Fairhead (2007) insightfully show how vaccination policy and practice today often enact a division between biomedical expertise (what they aptly refer to as "science-as-epidemiology"), and worried lay citizens whose actions are envisioned as based on feelings, personal experiences and misunderstandings of science. This, they stress, is often explained through the idea of a general decrease in citizens' trust in vaccination programs. In my study, I will make use of their notion of science-as-epidemiology, and how this becomes linked to, or contrasted with, feelings and trust.

HPV vaccines in STS and elsewhere

In relation to HPV vaccines, researchers have pointed toward matters both similar to and different from other vaccines. Notably, in the anthology *Three Shots at Prevention*, Keith Wailoo with colleagues (2010) discuss HPV vaccines as involving a "new politics of prevention" that centers around individualized risk rather than biopolitics governing populations *en masse*. In this collection of work it is emphasized that HPV vaccine politics articulate herd immunity[6] as something to be reached through

6. *Herd immunity* is a form of indirect protection from infectious disease. "When a critical portion of a community is immunized against a contagious disease, most members of the community are protected against that disease because there is little opportunity for an outbreak. Even those who are not eligible for certain vaccines—such as infants, pregnant women, or immunocompromised individuals—get some protection because the spread of contagious disease is contained" (US Department of Health 2016).

"one-by-one" population politics (Aronowitz 2010). Similar conclusions are drawn in other studies (see e.g. Polzer and Knabe 2009; Connell and Hunt 2010; Spratt et al. 2013; Vardeman-Winter 2012; Charles 2013, 2014; Davies and Burns 2014; Burns and Davies 2015).

Many studies show how HPV vaccines are part of gendered "anticipation regimes" that encourage girls (e.g. Mamo et al. 2010) and mothers (e.g. Reich 2014) to manage and calculate individualized cancer risk as a way of anticipating future health. These studies show how HPV vaccines might include a logic of futurity which asks girls and mothers to act *now* to safeguard a healthy *future*.

Despite the fact that HPV is not a gender-specific infection, by articulating HPV vaccines as vaccines against cervical cancer (Wailoo et al. 2010), HPV vaccines are enacted as directed toward the girl body (Casper and Carpenter 2008, 2009a, 2009b; Mishra and Graham 2012). As I pointed toward earlier in this chapter, a "cancer frame" (Epstein 2010; Epstein and Huff 2010; Lindén 2013b) has enabled a construct of HPV vaccines as gendered, girl-centered vaccines. By focusing on cervical cancer instead of HPV and sexual dimensions, HPV vaccines are constructed as vaccines for girls, and against cervical cancer. U.S. scholars emphasize that such "side-lining" of sexuality had to do with a politically conservative controversy staging HPV vaccines as allowing for "sexual promiscuity" (see e.g. Casper and Carpenter 2008). Similarly to how this previous research stresses how HPV vaccines are constructed as girls' vaccines through gendered assumptions, politics of gender, sexuality and girl-centeredness will be important matters in this study.

Many studies concern HPV vaccination campaigns (see Polzer and Knabe 2009; Connell and Hunt 2010; Vardeman-Winter 2012; Charles 2013, 2014; Davies and Burns 2014). In addition to how such campaigns, as shown above, articulate individualized risk, studies emphasize that HPV vaccination campaigns often rely on a postfeminist discourse of girl empowerment. For example, Christyn Davies and Kellie Burns (2014: 713) argue that a U.S. campaign for Gardasil "co-opted postfeminist tropes" of empowerment, this "in order to produce girls, young women, and their mothers [...] as agents of their own health".

The majority of research on HPV vaccines is from North America. However, Andrea Stöckl (2010) discusses the introduction of HPV vaccination in Germany, Italy and Austria in her chapter in *Three Shots at Prevention*. She emphasizes that whereas the focus in the U.S. has been on "a responsibility for moral lifestyle" (ibid.: 267), "the introduction of the HPV vaccine in Europe [is] largely focused on the relationship between the state and its citizens and on questions of transparency" (ibid.). Relatedly, in forthcoming work, a colleague and I (Lindén and Busse forthcoming) point toward a discursive transformation from a "girls' vaccine" into a "children's vaccine for everyone" as part of the introduction of HPV vaccines to boys in Austria. This focus, we argue, articulates changed relations between the individual, the population, and goals of herd immunity, which is framed through discursive claims about gender-neutrality. Additionally, Ilana Löwy (2010: 285–286) explains that the introduction of HPV vaccination in France was "one of absence – absence of public debate, of professional controversies, of real engagement with a public issue", and that this story, therefore, is strikingly differently from the North American ones. At the same time as these studies tell narratives different from those of North America, in a previous study on pharmaceutical company advertisements for Gardasil in Sweden, I discussed how articulations similar to the North American focus on girl empowerment and individual, yet girl-centered, responsibility, were also part of the Swedish HPV vaccine context (Lindén 2013b). Relatedly, Johanna Rivano Eckerdal (2015: 745, my translation from Swedish) discusses in a feminist STS-inspired study how the decision to include only girls in the Swedish vaccination program reflects "a common and criticized view presenting sexual and reproductive prevention as a woman's responsibility".

Outside the Global North context, some other researchers have studied HPV vaccination. For example, in looking at the introduction of HPV vaccines in Colombia, in his PhD thesis *Making Evidence, Making Legitimacy*, Oscar Javier Maldonado Castañeda (2015: 50) argues that, in Colombia, "HPV vaccines are simultaneously promoted and perceived as drugs for individual risk when they are distributed through the market and as public goods within government vaccination programmes". Notably, he shows that in Colombia HPV vaccination politics can be understood as including both global and local particularities. For example, differently from the U.S. and Sweden, in Colombian public health campaigns cervical cancer is presented "as a consequence of an uncontrolled sexuality and women as victims of men's promiscuity" (ibid.: 37). Similarly, Fouzieyha Towghi (2013: 334) stresses that the marketing of HPV vaccines in India can be understood as "global-local realignments" with "localized effects".

Health campaigns in STS and beyond

Humanities and social science researchers in Sweden and the other Nordic countries have done studies partly, or fully, on health campaigns. The majority of these studies are historical (see e.g. Olsson 1997; Torell 2002; Thorsén 2013), but there are also a few contemporary ones (see e.g. Johansen et al. 2013; Törrönen and Tryggvesson 2015). The majority of these uses a governmentality perspective, inspired by the work of Michel Foucault, and by later work conducted by Nikolas Rose and Peter Miller. From such a perspective, these studies emphasize how different "governmentalities" serve to regulate citizens, often through discourses of self-responsibilization and self-government.

Of special relevance for my study is work on campaigns related to sexual matters. One such study is a PhD thesis by David Thorsén (2013) that partly concerns state-financed Swedish AIDS campaigns between 1987 and 1996. He shows how a transformation in governmentalities has occurred over time. From being about "HIV and AIDS as something that could affect everyone" (Thorsén 2013: 290, my translation from Swedish), HIV/AIDS campaigns in Sweden became increasingly articulated as an individualized message (ibid.: 402).

Thorsén discusses how HIV/AIDS campaigns throughout the time period studied were focused on sex. Yet it has varied over time whether the main emphasis has been on sex as connected to risks (such as articulations of risks of having sex with members of specific, so called, "risk groups"), or if the message has "been more affirmative and openly positive toward sex" (Thorsén 2013: 330). Moreover, he emphasizes that campaigns directed toward teenagers "did not distance themselves from teenage sex" (ibid.: 298). At the same time, Thorsén shows that a more affirmative and positive approach to sex often has implied heteronormative assumptions about "good" sex as being a matter of monogamous sex between a man and a woman.

Anna Bredström draws related conclusions in her PhD thesis *Safe Sex, Unsafe Identities* (2008). From a feminist, intersectional perspective, she shows how a "positive view on sexuality" (Bredström 2008: 236) in Swedish HIV/AIDS campaigns (including campaigns directed at teenagers) came with exclusionary discourses that represented "risk groups" along gendered, sexual and "race" lines. Importantly, Thorsén and Bredström show how both a risk-oriented ("negative") *and* an affirmative and positive discourse around sex have reproduced exclusionary constructions of "risk identities".

Outside the realm of sexuality, several Swedish and Nordic studies have been conducted on state-funded health campaigns. One such example is the article "Why Take Chances?" (Leppo et al. 2014), in which alcohol health education campaigns in Sweden, Finland, Denmark and Norway are compared and analyzed. In all countries, the authors argue, the campaigns simplified and reduced complexities, and made uncertainties absent regarding risks from drinking during pregnancy.

Some Nordic studies emphasize how public health campaigns reproduce gendered assumptions. For example, Jukka Törrönen and Kalle Tryggvesson (2015) critically analyze campaigns addressing pregnant women about alcohol. Using a governmentality perspective, they show how the campaigns use emotional images and "fear-appeals" to convince mothers-to-be not to drink during pregnancy. The campaigns encourage mothers "to internalize a certain understanding of healthy and risky behavior" (ibid.: 74). This, they further emphasize, comes with gendered assumptions about mothers' care responsibility for others (their fetus). A similar focus on reproductions of gendered ideas concerning femininity is present in a Nordic study conducted by Venke Frederike Johansen with colleagues (2013). Through their analysis of breast-cancer campaigns, they illuminate that gendered stereotypes are reproduced. As many other researchers from outside the Nordic countries have also discussed (see e.g. Cartwright 1998; Wagner 2005; Jain 2013), they importantly highlight that such campaigns "pink-wash" cancer through gendered metaphors and symbols.

Yet another example of research on public health campaigns from Sweden is a study by Ylva Habel (2011, 2013), in which she uses a governmentality perspective to analyze a Swedish multi-medium campaign. During the summer of 1937 this campaign went on an extensive bus tour to promote milk as a tool to improve the health of individual citizens, and in its extension, of the nation. In drawing upon Foucault's notion "ethics of care for the self" (1988), Habel convincingly shows that "participatory strategies" figured as a way of governing citizens to care for themselves, and for the nation.[7]

7. Habel's study can fruitfully be linked to other historical studies that emphasize specificities regarding public health in Sweden. In history, such researchers highlight, public health initiatives have articulated a close link between citizens' individual (im)morality, and national prosperity and health (see e.g. Johannisson 1994; Palmblad and Eriksson 2014). This relied on the assumption of the individual as a part of the collective whole (population *en masse*) rather than as an unit separate from society.

Outside of the Nordic context, a few researchers working with STS approaches have taken an interest in public health campaigns. Among these, Pru Hobson-West (2003) looks at UK measles, mumps, and rubella (MMR) vaccination campaigns. Through a "language of choice, empowerment and individual responsibility in current public health discourse" (ibid.: 277), these campaigns presented the vaccination as parental choice, rather than a public duty.[8]

In the book *Making Disease, Making Citizens: The Politics of Hepatitis C*, Susanne Fraser and Kate Seear (2011) draw upon an actor–network theory (ANT) inspired perspective to argue for the multiplicity of hepatitis C in campaigns. In doing so, they conclude that health campaigns materialize hepatitis C as an object. Drawing upon the work of STS scholar Annemarie Mol (2002), they show that hepatitis C in this material is "more than one and less than many" (Fraser and Seear 2011: 44), and that it enacts both "self-care and care for others" (ibid.: 43). Through their analysis, they illuminate how such forms of care are multiple and shifting; care for the self and care for others are multiple phenomena.

Another example of an STS inspired study that emphasizes complexity is Roddey Reid's *Globalizing Tobacco Control* (2005), which concerns anti-smoking campaigns in California, France and Japan. Insightfully, Reid argues that such campaigns enact tensions between the specific and general, and the particular and universal. Attending to such tensions, Reid provides a nuanced narrative about campaigns as articulating what he conceptualizes as "global singularities". Helpfully for my study, he stresses that such processes include both inclusive and exclusive tendencies.

The anthology *Imagining Illness: Public Health and Visual Culture* (Serlin 2010a) is another example of an STS inspired work on public health campaigns. In the book's introduction, David Serlin (2010b: xxvi, emphasis in original) argues that public health campaigns are "shaped in large part not only by *what* they depict but also *how* and *where* they are depicted as well as in what contexts they initially emerge and to what contexts they ultimately flow". Reminiscent of, for example the studies by Thorsén and Habel, the chapters in *Imagining Health* show how the (material) media used in campaigns, as well their context, are vital aspects

8. The focus on individual responsibility and parental choice in contemporary vaccination campaigns can be contrasted with historical studies from Sweden (Axelsson 2004) and the U.S. (Durbach 2004; Colgrove 2006) that show how vaccination campaigns in the past simultaneously emphasized individual rights, state responsibilities and communal duties.

for understanding specificities and complexities of campaigns. Even if public health campaigns often can be understood as "what Foucault identified as a 'care of the self'" (Serlin 2010b: xxi), this collection of work helpfully shows that it is important to consider *how* and *why* this is made possible. This focus on specificities regarding context and medium is something that I bring with me to my study.

Outside of STS, a range of sociologists, media studies scholars, medical humanities researchers and critical public health scholars have taken an interest in health campaigns. As in the Nordic studies, they often use a governmentality perspective. In doing so, they emphasize how contemporary public health campaigns articulate a neoliberal discourse of self-management and individualized responsibility, and how they often reproduce different power differentials (see e.g. Lupton 1995; Tulloch and Lupton 1997; Fullagar 2002; Moulding 2007; Gagnon et al. 2010; Crawshaw 2012). For example, in the often-cited book *The Imperative of Health*, Deborah Lupton (1995) illuminates how public health campaigns govern citizens through gendered stereotypes and moralizing discourses about self-care. Importantly, Lupton (1995: 49) emphasizes that it is crucial to question "whose voices are being heard and privileged" in campaigns.

In another study, Lupton and a colleague (Tulloch and Lupton 1997) study HIV/AIDS campaigns. Through a thorough ethnographic, comparative and audience studies approach, they insightfully attend to the complexities and tensions in health campaign production, design and reception. They show that there is no straightforward and linear process between health promoters' production work, and publics meaning-making of health campaigns. Instead, they highlight that these parts are culturally produced, and related to diverse power relations, such as gender, sexuality, nationality and "race". In my study, I also make use of different methods to be able, as Tulloch and Lupton (1997), to attend to health campaigns as a multilayered phenomenon.

Some researchers discuss temporal dimensions of health campaigns. For example Rosalyn Diprose (2008) argues that Australian anti-smoking campaigns include a conservative image of the future. She stresses that the campaigns present a future-oriented temporality that assumes that "[i]t is better to be safe than sorry and preserve what is deemed good about the past that is still present" (Diprose 2008: 143). She emphasizes citizens' capacity to resist this conservative discourse in how their actions keep open an undetermined future (ibid.: 148). Relatedly, Qian Hui Tan

(2015) relates to discussions of the "geographies of the future" to stress citizens' capacity to resist a future-oriented anticipatory logic in tobacco-control campaigns. Another example is a study of obesity campaigns conducted by Rebecca Coleman (2015), in which she argues for the need to discuss *the politics* of futurity, meaning how articulations of futurity in campaigns engage people differently. She argues that how "the uncertainty of the future is brought into the present, is not felt or lived out in the same way by everyone" (ibid.: 185). Like Diprose, Tan and Coleman do, I problematize articulations of future-oriented time by pointing toward the politics of this temporality, and by emphasizing alternative temporal articulations.

Since my thesis studies digital media, I will end this research overview by bringing up some research of how digital media are used for health campaign purposes. As part of her bigger research program on digital health technologies, Lupton (2013: 8) studies digital media health campaigns. She argues that campaign methods using digital media such as apps and Facebook are imagined to "have the potential to far exceed the relatively blunt instrument of the social marketing campaign". Digital media campaigns, she argues, are represented as "facilitating 'engagement' of and fostering 'partnerships' with members of the public" (ibid.). This, she concludes, is used to encourage "members of targeted 'risk groups' to become responsible for promoting their own health" (ibid.). An empirical case study of digitally mediated health campaigns, is the one by Daniel Hunt and Nelya Koteyko (2015). They argue that participatory Facebook health information sites regulate citizens through a neoliberal discourse of "the self-governing, responsible and enterprising individual" (ibid.: 446). Importantly, they emphasize that Facebook simultaneously both transforms health campaign practice through enabling increased public engagement and reproduces predominant power relations. Similarly to Lupton, Hunt and Koteyko, I attend to digital media as participatory technologies that might transform relations between campaigns and publics.

From a different perspective, in "How to Have Social Media in an Invisible Pandemic", feminist visual culture and STS scholar Lisa Cartwright (2013) looks at the meeting between, what she defines as a slow and imperceptible temporality of hepatitis C and an immediate and fast temporality of social media, health campaigns and the H1H1 flu. She convincingly shows how focusing on empirical examples (such as hepatitis C) inhabiting a slow temporality, can help critique pervasive modes of futurity. Close to Cartwright's approach, I attend to different temporalities (such as slow and fast) as enabled and articulated by digital media health campaigns.

Relating my study to previous research

The research discussed here has illuminated a range of vital points for this study. As a general theme, many scholars emphasize that public health (including HPV vaccination and campaigns) in contemporary society often includes articulations and practices of individualized health, risk responsibility, and self-care. Helpfully, many researchers point toward the complexities and multiplicities of these practices by attending to coexistences of self-care and care for others, diverse temporalities and materialities, intersecting power differentials, generative tensions, and contextual and situated (national, local) singularities. These insights set an important basis for my study. In line with these researchers, I find it productive to hold on to the contextual and situated complexities of my case.

The research I discussed here effectively delineates how public health today is often a "care of the self" practice and sometime also a "care for others". In line with this work, in my own study, I have found it important to attend closely to articulations of such double matters of care. Articulations of both care for the self, and for others, are part of my material. At the same time, and as already stressed, I believe my material also allows me to say *other* things about care, things not confined to either a care for the self or a care for others. Thus, and while the emphasis on self-care and care for others in previous research provides highly valuable insights, this also sets the basis for an exploration of care that is not bounded by these matters of care. In other words, my empirical material enables an approach that attends to care in a context of health campaigns as something both multilayered, situated, material and temporal.

Outline of the study

In this chapter, I have introduced and contextualized the topic of my study, situated it in an empirical and theoretical setting, and explained its aims and research questions. In Chapter 2, I further introduce, and explain, the study's theoretical approach. I situate the study within feminist STS in general, and feminist STS care studies in particular. I contextualize the care STS conversation through its roots in, and affinity with, feminist ethics of care and feminist standpoint feminisms. Moreover, I explain how I attend to care as an ethico-political commitment, and as a matter of feelings, materialities and temporalities. Additionally, I unpack these aspects further by bringing in feminist affect studies, STS studies on absences, STS work on material devices, and STS and media studies research on media temporalities. Throughout the chapter, the theoretical notions used in this study are introduced and explained.

Chapter 3 discusses the methodological implications of my study's theoretical approach, and provides information about the empirical material. I describe the methods used: different forms of close readings combined with an STS device perspective and interviews. I end with some reflections on the politics of methods in STS.

Next I introduce the two first empirical chapters (Empirical Part I). I introduce the first "I love me" campaign, and the themes of its two empirical chapters. In the part's first empirical chapter (Chapter 4), the HPV app is discussed. The HPV app is attended to as inhabiting a vision of being a "care enabler" (Eidenskog 2015) that ought to facilitate girls' capacity to care for themselves through vaccination. I relate the app to other social studies and humanities studies on apps, and discuss how the HPV app is different from many other apps discussed. When trying to make sense of the specificities of the HPV app, I frame the analysis through a discussion around ethico-political dimensions of what Puig de Bellacasa (2011) discusses as a "care for neglected things", and through a focus on coexisting media temporalities.

In the second empirical chapter about the HPV app (Chapter 5), I attend to how my interviewees at Mittland County Council discussed the HPV app, and its relation to HPV vaccination information. I focus on a predominant articulation in the interview narratives: the need to immediately respond to vaccination fears and myths with proper, neutral scientific information to enable girls to choose. By highlighting frictions, uncertainties and alternative articulations, I stress moments in the interviews that disrupt and slow down a seemingly clear-cut vision of vaccine fears as something that needs to be counteracted *now*. The two chapters in Empirical Part I sets the basis for my focus on temporalities of care.

Then I introduce the two empirical chapters that center around the first "I love me" campaign. In this introduction (Empirical Part II), I also introduce the campaign. Chapter 6 is focused on the Facebook site as part of the first "I love me" campaign. By combining a focus on matters of care with an STS perspective on public engagement in digital settings,

I discuss how matters of care on the site were articulated through the, as I will conceptualize it as, "care collectives" of Bredland County Council, different publics and Facebook devices (such as the like and share social buttons). I attend to how Facebook devices on the "I love me" campaign site facilitated promotion of a campaign vision of "I love me" and HPV vaccination care as a matter of "anticipatory immediacy" (Puig de la Bellacasa 2015). Moreover, I attend to how devices also enabled a vaccination-critical public to trouble this vision, such as by allowing for articulations of feelings other than love, and other temporalities of care. I discuss this as a case of a vaccination controversy in which often conflicting matters of care were articulated.

The next empirical chapter (Chapter 7) draws upon interviews with the county council professionals from Bredland County Council about the first "I love me" campaign. In contrast to the previous chapter's focus on diverse publics on the Facebook site, in Chapter 7 I unpack the girl-centeredness of the interviews. I attend to different matters of care, such as girls caring for themselves, and the county council caring for girls. This chapter further investigates links between public participation and care, and temporalities and care. The chapter's focus on temporalities of care, which trouble visions of anticipatory immediacy, further develops the theoretical discussion in Empirical Part I.

Empirical Part III introduces the second "I love me" campaign, and the themes of its two empirical chapters. Chapter 8 engages in possibilities for responsible re-storytelling, by zooming in on the second "I love me" campaign's affective and temporal cancer narratives from ex-patients and relatives. In combining close reading with an implicated (auto-ethnographic) reading, I focus on multilayered dimensions of temporalities of care, as articulated by campaign cancer narratives, and by my engagement with the narratives.

In the last empirical chapter (Chapter 9), I discuss links between public accountability and care. This is done through interviews with professionals from Bredland County Council about the second "I love me" campaign. The chapter discusses boundaries drawn by the interviewees for when, and how, the campaign's focus on death and disease can be a responsible endeavor for the county council as a governmental agency. In doing so, I engage in a conversation concerning when, how, and for whom care can be a matter of public accountability. Empirical Part III further deepens the theoretical discussion through a focus on ethico-politics of care (temporalities).

In the concluding discussion (Chapter 10), I summarize my empirical and theoretical contribution, paying special attention to questions of public participation in health communication, and my work on the temporalities of care. I also discuss some of the questions my study has raised and the directions I see as important for future research on matters of care.

2. THEORY: Promises and Troubles of Care

In this chapter, I explain and define my theoretical approach. I will start by explaining a few analytical notions (material-semiotics, tropes, visions, mediations, re-presentations and articulations) that I use when I talk about the specificities of campaign (i.e. media) material. Then I will turn to the topic of care. To unpack the specificities in how STS researchers approach care, I start with a short overview of feminist standpoint theories concerning care and of ethics of care. From there I will turn to how I approach care as a matter of ethico-politics, feelings, materialities and temporalities. For clarity reasons, all the notions I use will be italicized on first mention.

Media as material-semiotics

I understand HPV vaccination campaigns as *material-semiotic* objects and practices. That is, I attend to how the campaigns include both textual, visual and material components, and how these things are relationally entangled and made. In doing so, I use a few material-semiotics *sensibilities* that help me attune the analysis toward such entanglements of materialities, visuality and text. That is, as a set of sensibilities material-semiotics starts from the assumption that social and material phenomena are made up of "webs of relations" (Law 2009: 141).

Inspired by Donna Haraway (2004b) I make use of the notion of *tropes*. I understand tropes as temporal configurations of figurative and metaphorical language. Haraway writes:

> Tropes swerve; they defer the literal, forever, if we are lucky; they make plain that to make sense we must always be ready to trip. Tropes are a way of swerving around. (Haraway 2004b: 2.)

Tropes "make us swerve, turn us around" and "[u]nless we swerve, we cannot communicate" (ibid. 2004c: 201). Formulated in this vein, tropes allow me to attend to media material as non-immediate and transformative. It becomes a matter of, as Haraway in the above quote formulates, swerving and troping. Attending to media and technoscience with such approach opens up possibilities for problematizing claims about a direct, unidirectional and immediate access to a world independent of tropes, of the metaphorical and figural. What is more, it makes it possible to attend to how not only materialities and semiotics are entangled, but also how figural tropes are involved in making worlds. Material-semiotics is about attuning to the *world-making* capacities of tropes. If "we can trope this world, we can – literally – make it swerve, make it turn" (Haraway 1997: 102). Tropes inhabit worlds, and worlds inhabit tropes. This is another notion I use: *inhabit*. It is a good metaphor for thinking about tropes as spatial and temporal phenomena that take part in world-making.

In this study, objects that involve visual components take up a lot of space: an app, Facebook, images, videos, and a trailer. The notion of *vision* is helpful here. It is a trope that, as Haraway discusses it (1991, 1997), points at the doing and technoscientific politics of sight. Using this notion allows me to ask questions about what visions inhabit, and what they do, enable and reproduce. That is, it makes it possible to attend to what these visual objects allow me and others to see, or not. Moreover, I attend to how visions are transformed and altered. Vision is therefore also a temporal trope that, notably, involves assumptions about possible futures.

A different way of approaching media material is to say that media is mediated, and is mediating things. As a temporal trope, the notion *mediation* serves as a critique of a vision of a direct, immediate relationship between representation and reality (for example, between girls represented in HPV vaccination campaign material, and girls "out there"). Attending to mediation is a way of troping worlds. A focus on processes of mediation serves a critique of a technoscientific vision of what Haraway (1991, 1997) defines as "a view from nowhere". A view from nowhere means to act without being seen, to represent without being represented. Hence, it is also a god-trick: seeing everything from nowhere. To problematize the idea of a view from nowhere, Haraway (1997: 190) uses photographs as an example. "There are no unmediated photographs [...] only highly specific visual possibilities, each with a wonderfully detailed, active, partial way of organizing worlds", she writes. In line with Haraway's approach, I pay close attention to the specificities in how visual objects (and other media objects) organize worlds. That is, what they *do*, and how this is made possible through mediation. I also attend to how things are made to look like they are not mediated; as views from nowhere.

In their book *Life After New Media*, feminist STS and media scholars Sarah Kember and Joanna Zylinska (2012) attend to mediation as a temporal matter. They introduce the notion of *temporalities of mediation* as a way of attending to how different media (such as apps, social media and posters) are relationally linked to each other. This means that different media are not distinct from each other. Instead, they are made through (webs of) relations to each other. Concretely, Facebook is not only social media, it is also, for example, images and videos that are uploaded on Facebook sites. Kember and Zylinska show how, for example, Facebook is enabled by a temporal entanglement, or coexistence, of different "older" media (e.g. images) and "newer" media (e.g. social networking social buttons). With this focus they show that it is important to consider the relational temporalities (flows) of mediation that enable different forms of media (e.g. images or Facebook).

A focus on temporalities of mediation enables an attention to media objects in terms of non-distinct relationalities. It opens up for an understanding of how "the content of media is always other media" (Kember and Zylinska 2012: 19), and that media objects do not carry any distinctive meaning in and of themselves. A focus on temporalities of mediation allows for attention to temporal processes of *remediation*, that is how digital media "refashions" older media (see Bolter and Grusin 1996). In different ways, in Chapters 4, 6 and 8 I attend to temporalities of mediation.

I use the word representation, as it is a helpful word for attending to visual and textual specificities of campaign material. For example, I discuss campaign images of girls and other subjects as representations. However, in the light of the ongoing feminist theory critique stating that this notion is reproducing a "representationalist" idea of reality (an unmediated, immediate relationship between representation and reality), I will throughout write *re-presentation* instead of representation. Like Karen Barad (2007: 133), I do not believe words (and visuals) have "the power [...] to mirror preexisting phenomena". Instead, they take part in making those very phenomena. As STS scholar Michael Lynch (2014: 324) pays attention to, re-presentations include a temporal dimension: "re-presentation, presenting again (and again and again, indefinitely)". This neatly highlights how re-presentations do not mirror preexisting phenomena. Instead, re-presentations are made up of temporal acts of staging, and re-staging. Therefore, I focus on re-presentations as a phenomenon of transformations, movements and, consequently, as being part of temporalities of mediation.

Re-presentations (and re-presentational objects) are certainly not everything. And as such, their specificities need to be attended to. This includes looking at how they are entangled with other material-semiotic elements. It is also important to pay attention to what re-presentations do in situations where people claim that what they are doing are in fact mirrorings of a given, external, and preexisting reality. Thus, in its specificity, the notion of re-presentation is useful for understanding a vision of re-presentational objects as enabling direct access to an external reality. That is, it is helpful for attending to "representationalism" as a pervasive, world-making, trope (for similar arguments, see Cartwright 2014; Coopmans et al. 2014).

Finally, I use the notion *articulation* as yet another temporal trope. For example, re-representations such as campaign images are brought into existence through acts of articulation. I understand articulation as material-semiotics, and not as simply a matter of language or discourse. Articulations, according to Haraway (1992, 1997), are about making linkages between heterogeneous elements, and as such they are about world-making. Thus, articulation sets things in motion, and align them with each other, and the notion can be used to denote how technoscience and media are about swerving. Haraway writes:

> Discourse is only one process of articulation. An articulated world has an undecidable number of modes and sites where connections can be made [...] To articulate is to signify. It is to put things together, scary things, risky things, contingent things. (Haraway 1992: 325.)

In pointing toward articulations as acts of making linkages between different elements, Haraway's definition of articulations concerns temporalities and contingency. Following this, I understand articulation as acts of setting in motion, to enable different ("risky") connections, to facilitate things capacities to swerve. Thus, articulation is a highly worldly and figural practice in how it involves textual, visual and material components that take part in making worlds. This makes articulations as an analytical notion helpful for my study since it allows me to attend to the making of temporal connections between diverse elements (humans, digital technologies, visuals, texts etc.). For example, I discuss how articulations enable linkages between publics and Facebook devices on the "I love me" Facebook site, and how this facilitates different temporalities of care.

An overview of care research: ethics of care, feminism and STS

Care is not a new topic in feminist theory. In the late 1970s and early 1980s, psychological theorist and ethicist Carol Gilligan (1977, 1982) argued that female subjects have a "different voice" than males. She stated that differences between males and females "arise in a social context where factors of social status and power combine with reproductive biology" (Gilligan 1982: 2). She emphasized that rather than being focused on general and objective principles about justice, women's morals are connected to interdependency, care, responsibility and empathy. A feminine voice, she stressed, is focused on interpersonal relations; it includes a close connection "between conceptions of the self and conceptions of morality" (ibid. 1977: 516). Even if the gender essentializing tendencies of her approach are often critiqued, Gilligan's sensibility toward care as devalued and feminized, and as denoting interdependence, relationality and responsibility toward others is an important basis for work in feminist ethics of care (see e.g. Tronto 1993; Sevenhuijsen 1998, 2003; Fine 2007).

More explicitly, feminist ethics of care researchers who relate to Gilligan's work emphasize relationality, interdependence, responsiveness, responsibility and attentiveness. Thus, care is often discussed as simultaneously a social and political activity, and a moral orientation toward others. Notably, Joan Tronto (1993: 118–119) defines care as a practice and a moral disposition. In line with this focus on care, in ethics of care, questions of care responsibilities or obligations are often emphasized. Some researchers stress specific obligations to care for special relations (Kittay 1999: 53–58), while others prefer "a flexible notion of responsibility" (Tronto 1993: 133).

Responsibilities or obligations to care are often formulated as closely linked to a need for responsiveness toward others. Responsiveness, Tronto (1993: 136) argues, suggests that we are "engaged from the standpoint of the other, but not simply by presuming that the other is exactly like the self". Relatedly, Selma Sevenhuijsen (2003: 186) stresses the importance of willingness and capability to attend to "the perceptions and viewpoint of others", without assuming the possibility of putting oneself in someone else's shoes. She stresses this "implies careful and respectful listening and responding" (ibid.: 187) to others. As I will discuss, this focus on responsiveness, capability and willingness is important in feminist STS approaches to care, and are also themes that I will attend to in this study.

Feminist researchers emphasize that care is marginalized and devalued in contemporary society and science. In this vein, feminist science studies scholar and standpoint theorist Hilary Rose (1983, 1994) argues for the need of a feminist science that starts from women's everyday material and bodily experiences of care labor. A "thinking from caring" will, according to Rose, enable a feminist, responsible science. Relatedly, Tronto (1993) and Sevenhuijsen (1998) stress that care is a vital life-sustaining activity that has the capacity to transform societal and political matters into the better.

It is clear that feminist ethics of care researchers and standpoint feminists have from the start embraced a political vision that sees care as central for societal transformations. Importantly, this involves an attention both to the problems of care (such as how care is feminized), and to its potential to transform society, science and technology.[1]

Firmly rooted in the conversations discussed above, care has been taken up in STS in a manner that is partly reminiscent of how care has been, and is, discussed elsewhere. Thus, STS care studies draw upon feminist ethics of care (see e.g. Mol 2008; Mol et al. 2010a), and feminist standpoint feminists such as Rose (see e.g. Puig de la Bellacasa 2011; Martin et al. 2015). Notably, in line with feminist ethics of care, STS researchers emphasize care as a matter of relating to others, and as denoting attentiveness, responsiveness and particularities rather than general

---

1. Researchers working with ethics of care sensibilities today are from a wide range of areas, including disability studies (Hughes et al. 2005; Thelen 2015); postcolonial studies (Newstead 2009; Raghuram et al. 2009) and geography (McEwan and Goodman 2010; Milligan and Wiles 2010; Bowlby 2012). Building upon "traditional" ethics of care scholars, this heterogeneous body of work simultaneously points toward care as a practice that is gendered and that comes with exclusions, and as involving an ethical and political potential and promise for more caring relations and societies. Care is discussed as an affective and bodily process, and as an ethical response, relating toward other humans and nonhuman animals, and to societal happenings (e.g. McEwan and Goodman 2010; Thelen 2015). Several of these researchers point toward care as something more-than-human by emphasizing, for example, material spaces such as museums and homes as facilitating care (Bowlby 2012; Munro 2013), and a care for "the environment" and nonhuman animals (Miele and Evans 2010).

ethical judgments. Moreover, in line with an ethics of care emphasis on care as a practice and as an everyday material experience, it is emphasized that care is a practice. In this vein, feminist STS scholar Puig de la Bellacasa (2011) draws upon the work of Tronto (1993) and Rose (1983, 1994) to problematize how care is being marginalized in contemporary society, and argues that it is an ethical and political move to try to foster more caring standpoints. Similarly, Annemarie Mol with colleagues (2010b: 13), stress that they build upon the ethics of care focus on "local solutions to specific problems", rather than general ethical principles.

Working with an ethics of care sensibility toward care as a material and relational practice, STS scholars work empirically to pay close attention to material specificities of care. Through a focus on care as a socio-material and/or material-semiotic practice, it is stressed how care is enacted through more-than-human elements and arrangements, including materialities such as technologies and devices (e.g. Singleton and Law 2013) and humans and nonhuman animals (e.g. Singleton 2010, 2012). Moreover, STS care studies clearly stress the ethics of care as a practical doing rather than as a moral stance. Ethics is being made through "situated, complex everyday practical entanglements of matter and ethics" (Latimer and Puig de la Bellacasa 2013: 155). Accordingly, Puig de la Bellacasa (2011: 95) emphasizes that it is dangerous to idealize care and turn it "into a fairly empty normative stance disconnected from its critical signification of a laborious and devalued material doing". Relatedly, Mol with colleagues (2010b: 13) stress that care is not something to be judged "in general terms and from the outside, but something to do, in practice". Thus for these STS researchers, care is about being attentive toward the specificities of socio-material practices of care. In this vein, STS care studies do not aim to "go beyond" ethics of care, but try to develop further the focus on materialities, specificities and practices of care.

Care in this study: ethico-politics, feelings, materialities and temporalities

This overview of care research sets the basis for my study. I situate what I am doing as in close affinity with feminist STS approaches to care (e.g. Singleton 2010, 2012; Puig de la Bellacasa 2011; Martin et al. 2015). Therefore, it is also in affinity with theoretical and political sensibilities drawn from feminist ethics of care and feminist standpoint feminisms.

Working with care in this spirit, I emphasize that I think from somewhere, and with others. I am anything but "from nowhere".

I attend to care as both a practice and a standpoint. This signifies two dimensions of care that from the very start of feminist ethics of care have been closely interlinked. The first dimension is the care the actors (humans and nonhumans) studied articulate and do, and the second is the things I care for as a researcher, and want to foster. For example, similarly to Tronto's (1993) attention to care as a standpoint concerning one's responsibility toward others and as an everyday practice, this means that care is always, already both an empirical finding and a political ethos.

Drawing upon Puig de la Bellacasa (2011), I use the notion of *matters of care*. She introduces the notion by reworking STS scholar Bruno Latour's (2004, 2005) notion *matters of concern*.[2] Compared to matters of concern, the notion matters of care, she argues, "has stronger affective and ethical connotations [and] is more easily turned into a verb: *to care*" (Puig de la Bellacasa 2011: 89–90, emphasis in original). Whereas care comes with connotations of "attachment and commitment", concern "denotes worries and thoughtfulness" (ibid.: 89).[3]

For Puig de la Bellacasa (2011: 90), care signifies "an affective state, a material vital doing, and an ethico-political obligation". I work with this definition of care as a heuristic. As such, it is what Haraway (2004a: 335) calls a "thinking technology"; something to *think with* to sensitize myself toward what care might signify. Thus, Puig de la Bellacasa's definition works as navigation for *how* I look for care. I will use her definition to talk about the *ethico-politics*, *feelings* and *materialities* of care. However, how to care requires different approaches in different situations, and a responsiveness to the situated specificities of care is required. Thus, I try to approach care with curiosity and responsiveness toward its particular material-semiotics.

---

2. *Matters of concern* is a notion invented by Latour (2005). He argues for it as a response toward deconstructionist approaches that use ready-made explanations for "debunking" relations of power and illusions. He uses the notion to stress the importance of constructionist approaches that focus on the material making of science and technology, and of facts. He asks for an approach that tries to protect and care. In an ANT spirit, he stresses that "[t]he critic is not the one who debunks, but the one who assembles" (ibid.: 246). Latour's focus on assembling matters and attending to care instead of an impetus of debunking, Puig de la Bellacasa (2011) develops further through the notion of matters of care.

3. In Chapters 5 and 6 both matters of concern and matters of care will be used as analytical notions. This will be further discussed in these chapters.

As discussed in Chapter 1, I also attend to *temporalities of care*. In doing so, I draw upon work from scholars discussing connections between care and time (see e.g. Haraway 2011; Puig de la Bellacasa 2015; Schrader 2015; Felt 2016), and, when necessarily, I also relate to other STS and feminist theory discussions on temporalities. Since temporalities of care in my empirical material are linked to ethico-politics, feelings and materialities of care, I will start by focusing on these parts of care, and then expand upon this through a focus on temporalities. As I hope will be clearer throughout this chapter, the reasons for why I focus on these components of care is simultaneously rooted in empirical findings, and in ethico-political commitments. In this vein, it is due to the two dimensions of care already discussed: "my" care and the care circulating among the actors I study.

I aim to allow the human and nonhuman actors inhabiting my material to help defining what care is in different situations. This approach makes it possible for them to teach me about care. As do Joanna Latimer and Puig de la Bellacasa (2013: 169), I work with an ethico-political commitment to care as a mutual constitution, instead of adopting "the role of an arbiter pointing out the right and wrong". I try to take seriously (and learn from) what the actors in my empirical material care about. In doing so, I attend to the multilayered dimensions of their care, such as what it enables, leaves out and marginalizes.

To attune to the ethico-politics, feelings, materialities and temporalities of care, I center my analysis around two figural, world-making tropes already introduced in Chapter 1: care as a *promise* and care as a *trouble*. This makes it possible to attend to the complexities of care. More care is not automatically something desirable. There is a much harmful and exclusionary care happening. For example, and as I discussed in Chapter 1, health campaigns, including those for HPV vaccination, frequently involve a gendering and moralizing of care. By articulating that people *should* care for themselves by adopting normative constructions of health such campaigns include exclusionary and normative articulations of care. Thus, care does not stand outside an already troubled world; it does not offer an escape (Murphy 2015: 732). Care is political and non-innocent (Viseu 2015; Giraud and Hollin forthcoming).

As others argue, the political and non-innocent dimensions of care make it important to "stay with the trouble" (Haraway 2010; Atkinson-Graham et al. 2015; Martin et al. 2015; Leem 2016). To stay with the trouble, in my study means holding on to the complex ethico-politics involved in matters of care. To allow for this, I try to slow down the analysis (Haraway 2008: 83; Jerak-Zuiderent 2013, 2014) instead of "brushing over" complexities.

When attending to care, I thus partly "look for trouble" (Roberts 2015: 37). To trouble a specific articulation is to aim for loosening "some of the tighter knots" (ibid.: 31) that hold it in place. It is about "the performative aspects of ethical and political imperatives that undo, trouble, keep open and push boundaries" (Latimer and Skeggs 2011: 408). Thus, to stay with care as a trouble is to keep opening up the layers (or knots) that hold stratified worlds in place. It is to move and unsettle things; to facilitate them to swerve.

I also look for promises of care, and care as involving promising moments and connotations. As others do, I emphasize that attending to care is promising in how it can transform material-semiotic relations, and create more livable worlds (see e.g. Tronto 1993: 122; Haraway 2011). However, I also attend to promises as involving specific visions of futures. As Ahmed (2010: 30) writes, promises indicate "something favorable to come". Writing on happiness as a promise, this "something", she explains, is conditioned on action. The idea is that "if you do this or if you have that, then happiness will follow" (Ahmed 2007: 12). That HPV vaccination campaigns promise things (like a cervical cancer free future) is not a surprise. However, as I will show, such promises are often anything but straightforward. For example, in Chapter 6 I will attend to how care changes when different actors trouble a promissory campaign vision. Following this, care as involving promises is something multilayered and ambivalent.

Altogether, by drawing upon feminist scholars who stress the ethico-political potential in attending to care *and* the problematic ways in how care is being done, I will attend to care as inhabiting multilayered promises and troubles. These scholars teach me that attending to care involves no straightforward guarantee for more livable worlds, and for practices of caring well. But, as will be focused on in the analytical chapters, focusing on promises and troubles of care facilitate an analysis that unsettles, slows down and pushes HPV vaccination campaign tropes and worlds. This focus allows me to pay close attention to the complexities of HPV vaccination campaigns, and their actors and politics.

Ethico-politics of care: a commitment to specificities, absences and marginalizations

I work with an ethico-political commitment where I try to enable and foster more caring relations, and trouble exclusionary and moralizing ways in how HPV vaccination campaigns are articulated. As others have pointed out, this turns care into a way of trying to *respond* responsibly toward other humans and nonhumans, and to political matters in society (Ahmed 2010: 186–187, 222–234; Atkinson-Graham et al. 2015: 739–741; Martin et al. 2015; Schrader 2015).

In my analysis I try to respond responsibly by paying close attention to both promises and troubles of care. This requires me to foster an openness toward how I get affected by the human and nonhuman actors I study, and how I affect them. It requires an *ethico-politics of response.* Trying to take seriously what my interviewees care for is one way of responding carefully. Similarly, I try to "stay with" the campaign material to allow it to surprise me, and as I have already said, teach me about care. Thus, I try to relate to my material with curiosity since this allows the actors I study to change my convictions and pre-understandings.

As a way of responding carefully, I work with feminist theory and STS sensibilities toward the political significance of the marginal and absent (see e.g. Puig de la Bellacasa 2011, 2014; Rappert 2015). In doing so, I explore what it might mean to hold on to, and care for, the marginal, neglected and absent in my study. As will be explained further in Chapter 3, I do not merely look for the big picture or for predominant articulations. Instead, I work with different sensibilities for attending to how small details, marginal things and subtleties can trouble and alter the predominant. For example, I discuss how hesitations and precautionary wordings in interviews make present other matters of care than those involving a "clear-cut solution" to what the actors I have talked with often articulated as a major concern: girls' fear of vaccinations. In general, in the parts of the study where I focus on the marginal and neglected, I aim to tell "alter-narratives" that can trouble predominant articulations. Inspired by Puig de la Bellacasa (2011, 2014) I emphasize a care for the neglected, marginal and absent as an ethico-political commitment that can foster alternative visions, and enable other worlds.

In focusing on absences, I work with what STS scholar Brian Rappert (2015: 21) emphasizes as "sensitivities for attending to what is absent". I do not believe it is generative or responsible to attend to absences as an act of simply adding up the absent voices. As Puig de la Bellacasa (2014: 37) reminds me, "[w]hat might seem absent from one practice's perspective is at the core of another's focus". Instead of being a simple act of adding the missing pieces, presences and absences are dynamically intertwined; presences depend on absences (Law and Singleton 2005: 342; Rappert and Bauchspies 2014: 2). It is all about *for whom* something is absent or marginal, and what politics is involved in this. Thus, making things absent is not something neutral, but neither is making it present. In line with this, in Chapter 4 I ask questions about for whom a care for neglected things can be important, and for whom such care perhaps rather involves troubling and exclusionary dimensions.

Feelings of care: promising and troubling happiness and unhappiness

As I brought up in Chapter 1, health campaigns, vaccinations and cancers are affective phenomena. It makes sense to talk about these things as involving affective matters of care. However, what this means is not a given and can be conceptualized in a range of ways. Care is easily equalized with "warm" feelings of love and nurture. As feminist STS scholars recently have highlighted, such a conceptualization of care can be problematic in how it sees care as something intrinsically good and positive. For example, Michelle Murphy (2015) draws upon the work of Sara Ahmed (2010) to caution against a tendency to equate care with positive or happy feelings (or as she calls it, affects) that easily can be imagined as opening up for, or affirming, a more promising world.

It is problematic to ignore dimensions of care that have to do with unhappy feelings such as worry, indignation, discomfort and pain. In my empirical material many unhappy feelings are articulated, and linked to matters of care. Vaccine fears were one of the main concerns for my interviewees at Mittland County Council, and on the "I love me" campaign Facebook site different publics articulated a vide register of unhappy feelings as critical responses to the specific re-presentations of care *as* love and happiness in the campaign material related to the first "I love me" campaign. I also stage my own discomfort toward the second "I love me" campaign campaign's affective capacities as an opening for retelling the campaign cancer narratives responsibly. I work with an unhappy feeling as a way of engaging matters of care as an ethico-political commitment.

In unpacking feelings of care further, and situating myself, I turn to feminist discussions on feelings and affects. I work with affect theories that can be labeled "cultural politics of feelings" (see e.g. Ahmed 2004, 2010). In choosing this label and this standpoint, I emphasize the "trouble" with feelings; they are evoked and circulating within stratified worlds. In this way, I attend to feelings as conditional and contextual, including how they sometimes reproduce dominant social and political exclusions and hierarchies. That is, I partly look for how feelings might *close down* possibilities for transformation, and reproduce predominant, and often exclusionary and normative, articulations. Feelings are linked in a range of ways to, for instance, normative doings of gender, "race" and sexuality (see e.g. Hemmings 2005; Pedwell and Whitehead 2012). In line with this, I focus on how articulations and re-presentations of love and happiness in my material involve a gendering of girlhoods, and a normative vision of what a good future for girls should entail.

I also look for moments where feelings *open up* for transformation and where they trouble and unsettle, rather than reproduce, hierarchies and exclusions. Feelings – may they be happy, ugly and/or embarrassing – do things. I do not presume that "bad feelings are backward and conservative and good feelings are forward and progressive [and] that good feelings are open and bad feelings are closed" (Ahmed 2010: 216–217). Instead, in this study, a wide range of feelings are attended to as sometimes opening up for more caring and inclusive relations, and sometimes as reproducing exclusionary and normative articulations of how one should live life. In attending to both "closed" and "open" feelings, I focus partly on how feelings orient people; sometimes they serve to *align* people with other subjects and with (more-than-human) collectives (and, as I will discuss, publics), and with ideals and material things. Other times they generate acts of *distancing* from other human subjects, and from material things (see Ahmed 2004, 2010).

Materialities of care: HPV vaccination campaign care devices

I focus on *care devices* (see e.g. Pols 2010; Singleton and Law 2013; Leem 2016) as a way of taking seriously materialities of care. I conceptualize a range of things as care devices: Facebook social buttons such as the like and share buttons, the HPV app, campaign evaluations brought up and used in interviews, and the "I love me" trailer are all approached as care devices. I define them as *care* devices as they participate in articulating care.

Drawing upon STS scholars John Law and Evelyn Ruppert (2013: 230), I attend to how "devices assemble and arrange the world in specific social and material patterns". As they also write, devices are materially heterogeneous:

> Devices may and often do include pieces of kit. More generally, patterned arrangements include materials that may but do not need to be high-tech (clipboards and pencils are just as material as nuclear reactors, radiation monitors or computers). But they are heterogeneous because they (usually) include people too (technicians, evacuees, electricity users, members of the technocratic middle class, interviewees). We might add that they typically include texts, inscriptions, representations or symbols too (sample statistics, temperature readings and graphs revealing radiation spikes). (Law and Ruppert 2013: 231.)

In a vein similar to Law and Ruppert, I attend for example to the HPV app and the Facebook social buttons as devices that assemble relations, narratives and worlds. That is, I focus on what these devices *do*, socially and materially.

STS scholars working on care practices such as telecare (Pols 2010) and cattle farming (Singleton and Law 2013) have shown that care devices articulate/enact specific versions or forms of care. Such care devices can, for example, "make invisible care work visible" (Leem 2016: 38) or generate feelings such as happiness and love. I attend to these device-mediated doings of care: I discuss how "care devices" can make neglected or absent things present, and how they generate and facilitate diverse feelings. I also attend to care devices as, what Maria Eidenskog (2015) defines as, *care enablers*. That is, how devices distribute and facilitate other actors' capability to care. For example, the HPV app can be discussed as a care enabler: it is envisioned to facilitate girls' capacity to care for themselves through vaccination.

Temporalities of care: troubling anticipatory, immediate and linear time

So far I have discussed the ethico-politics, feelings and materialities of care. Focus has also been put on how I attend to care through different temporal tropes: temporalities of mediation, visions, re-presentations and articulations. However, I have not explicitly unpacked how this focus on temporalities relates to matters of care.

In the campaigns I study, girls are often re-presented as needing to anticipate future cervical cancer by taking care of themselves now (by getting vaccinated). The campaign material often involves articulations of "anticipatory time" that poses the future as urgent and immediate; as something that needs to be acted upon now. As Vincanne Adams with colleagues (2009) have discussed, such anticipatory time often works as a normative affective state that asks subjects, such as girls, to calculate fears and hopes through articulations of a moral (and gendered) responsibility for future health.

At the same time, anticipatory and immediate temporalities of care are not the only ones articulated in my empirical material. Instead, the material involves a range of actors and moments that complicate and trouble such temporalities of care. For example, I will discuss how Facebook devices on the "I love me" Facebook site participated in articulating alternatives that redirected and slowed down anticipatory temporalities of care. As an ethico-political commitment, I pay close attention to such troublings of anticipatory and immediate time.

Several STS and feminist researchers have discussed connections between care and time (see Tronto 2003; Haraway 2011; Kember and Zylinska 2012: chap. 6; Lappé 2014; Felder and Oechsner 2015; Puig de la Bellacasa 2015; Schrader 2015; Felt 2016: 193). They discuss how care requires and/or fosters a reorientation of time. Reorient here denotes a critique and problematization of technoscientific visions and practices of linear, progressive, productivist, teleological, immediate, goal-oriented time and/or anticipatory time. Such time is often discussed as a mode of technoscientific futurity; a future-oriented vision that equals the future with progress, and the present with actions directly leading to the realization of that future.

Helpfully, feminist work (Haraway 2010, 2011; Puig de la Bellacasa 2015; Schrader 2015) illuminates how attending to matters of care can complicate and disrupt visions of immediate and anticipatory time. In a similar vein, I aim for telling time in a manner that allows for disruptions, for example, of calls for urgency and immediate action (e.g. "get vaccinated now!"). Helpfully for this endeavor, Astrid Schrader (2015) discusses possibilities of a less anthropocentric time that does not articulate care in terms of a demand for progressive, immediate and direct action. In turn, Puig de la Bellacasa (2015: 704–705) focuses on alternative, as she calls it, "care times" that disrupt "the productivist futurity dominating contemporary technoscience". This, she argues, involves attention to a multiplicity of entangled timelines. In different ways, Schrader and Puig de la Bellacasa show how an attention to multiple timelines can trouble calls for linear, anticipatory and immediate time. They help me disrupt articulations stating that girls (and others) need to care now to safeguard happy futures.

As I have already partly mentioned, one way I concretely aim for disrupting visions of future-oriented and immediate time is by engaging matters of care that slow down calls for urgency. By relating to other scholars (Haraway 2008: 83; Jerak-Zuiderent 2013, 2015; Martin et al. 2015: 658; Schrader 2015: 638) who attend to how a slowing down of time might open up for an alternative, and more caring, engagement, I attend to moments in the interviews and the campaigns that inhabit or enable a slower temporality of care. I will discuss moments and situations that allow for alternative and more caring spaces and times, or as I will discuss it as *space-times* (Schrader 2015). This means that engaging alternative temporalities of care is also an ethico-politics of response. Therefore, I respond to calls in my empirical material for progressive and immediate action by staging other temporal engagements that make possible alternative visions, narratives and worlds.

In summing up, I focus on multiple and coexisting temporalities of care. Some of them reproduce visions of anticipatory and immediate time, others disrupt such visions, for example by slowing time down. I also attend to temporalities of care as an ethico-political commitment where I try to disrupt normative future-oriented visions that privilege "the new". Moreover, attention is put on links between temporalities of care and materialities of care (especially through attention to how devices mediate temporalities of care), and to links between time and happy and unhappy feelings such as through promises of happy futures.

3. METHODOLOGY: Studying HPV Vaccination Campaigns with an Ethnographic Attitude

Matters of care are simultaneously about theory and methodology. In this chapter, I discuss the *methodological implications* of my theoretical approach. Moreover, I provide information about my empirical material: the campaign material and the interviews. Additionally, I introduce, and explain, my methods: close readings combined with an STS device perspective, and interviews with professionals.

I commit myself to what Haraway (1997: 190–191) defines as an "ethnographic attitude". She explains this as:

> [A] method of being at risk in the face of the practices and discourses into which one inquires [...] An "ethnographic attitude" can be adopted within any kind of inquiry, including textual analysis [...] [A]n ethnographic attitude is a mode of practical and theoretical attention, a way of remaining mindful and accountable. Such method is not about "taking sides" in a predetermined way. But it is about risks, purposes, and hopes – one's and others' embedded in knowledge practices [...] [It] is a collective undertaking that cultivates a practice of learning to be at risk. (Haraway 1997: 190–191.)

For Haraway, an ethnographic attitude is not confined to doing fieldwork *in situ*. In fact, her main research materials are a combination of texts and visuals, and stories from her own life (like dog agility training; see Haraway 2008). Instead of being confined to a specific method, an ethnographical attitude is, according to her, about putting oneself at risk (one's subjectivity, assumptions, views, etc.) in the meeting with others (humans and nonhumans), and remaining careful and accountable when doing so. Thus, it is a mode of attention, a matter of practical doing, and a relating. Learning to be at risk is to "challenge previous stabilities, convictions, or ways of being of many kinds" (Haraway 1997: 191), and trying to do so in an ethical and responsive manner. In line with this, Beverly Skeggs (1999: 48) translates Haraway's ethnographic commitment into "an ethics of witnessing which is both responsive to and responsible for". Formulated in this way, an ethnographic attitude is in close affinity with an attention to care, as an ethico-political commitment, and as a matter of learning to respond carefully. It becomes a matter of accounting for how one is affected, moved and transformed by the study (and simultaneously, how one affects it).

Translated to this study, I engage an ethnographic attitude to *learn about* care from the diverse human and nonhuman actors in my material, instead of "taking sides" in a predetermined way. Thus, I want to allow the actors in my material to challenge and unsettle my convictions and views, and I try to account for how I, in turn, move and affect them.

My empirical material is diverse. It consists of non-digital and digital images, videos, a "HPV app", feedback messages from girls sent via the app to one of the county councils, a Facebook site involving other county councils, diverse publics and digital devices, written campaign cancer narratives, auto-ethnographic writings, as well as interviews with health care and health administration professionals working with the campaigns. Thinking about this set of material in terms of an ethnographic attitude is helpful at it takes seriously the generative connections articulated when diverse materials are allowed to constitute an object of study. That is, it makes possible a multilayered approach. Before I move on to discuss how this is played out in relation to the different methods I work with, I need to present my material.

The campaign material: the HPV app and the two "I love me" campaigns

The campaign material from Mittland County Council consists of the app and 40 feedback messages sent to the county council from users via the app. It also consists of a cinema advertisement, an information poster and a newspaper advertisement about the app. These latter entities are not analyzed in this study, as I have decided to focus specifically on the digital media objects. At Mittland County Council, the HPV app was decided on based of the findings from a set of focus group interviews. There the county council interviewed high school girls about how they believed girls today want to, and can, be reached with health information. These focus groups are not directly part of my empirical material but were brought up in the interviews.

The campaign material in relation to the two "I love me" campaigns is larger and more diverse. The first "I love me" campaign consisted of six campaign periods. In February 2012, an information pamphlet was sent to young women (between 18 and 20 years old) or to guardians (for girls under 18 years old), two posters were located on public transport, and an advertisement was printed in three different local newspapers. Following this, in May 2012, campaign material was sent to high schools in the Bredland region. It consisted of a pamphlet, a reminder card, a health declaration paper, a covering letter, and two posters. In June 2012, a summer campaign was launched that consisted of three posters located in the public transport system, a postcard sent to girls, a postcard sent to parents and an advertisement in a newspaper. In August 2012, a campaign event at a regional high school fair was carried out. This included the possibility of getting vaccinated at the location in the vaccination trailer. It also included four placards located by the showcase, the same information pamphlet and reminder card as was used during previous campaign periods, and information uploaded on the high school fair's webpage.

During the fair, canvas bags with the "I love me" logo were given out to fair visitors. Following this, during the fall of 2012, an extensive school tour was conducted with the possibility of vaccination in schools, in the vaccination trailer. The tour will here be discussed as the "vaccination trailer tour". Before the vaccination trailer stopped by a specific school on the tour, five posters, a cover letter, health declaration and the information pamphlet previously used were sent to the schools. An advertisement was posted in one newspaper, and digital campaign images and texts were posted on the high schools' own homepages. During the tour, the "I love me" canvas bag was given out. In connection with the trailer tour, information posters and signs were placed near the trailer. Additionally, during most of the first "I love me" campaign's existence, it included extensive material posted on the *Care Guide* web page[1], and on a campaign Facebook site. In total, not counting the information on the *Care Guide* and Facebook sites, this material consisted of approximately 50 diverse entities (images, a trailer, a bag, pamphlets, letters, posters, etc.).

The Facebook site was launched in March 2012, and ran until the spring of 2013. The majority of the campaign images used in other settings were uploaded to this site. The Facebook site also consisted of extensive additional visual and textual material (both material designed

1. *The Care Guide* is a Swedish web page that provides information about diseases and health, as well as about health care provision in Sweden.

specially for the "I love me" campaign, and other images). The county council documented the high school vaccination trailer tour through images (around 40 images) and videos (16 videos) that were uploaded to the Facebook site. In addition, the Facebook site generated hundreds of status updates (including textual and visual material), comments, likes and shares (from the county council and different public users).

I have downloaded the images and videos from the Facebook site. The whole body of Facebook material was collected by me through screenshots (537 screenshots). I accessed the Facebook material after the site was publicly closed down. To enable access, Bredland County Council added me as a site administrator for the site between February and June 2014. Since the site was at that time still online, but not publicly visible, this enabled me to access all the material (including the Facebook conversations between users and the county council that took place) on the site during its existence. However, since the site was not any longer publicly running, I could not follow *ongoing* conversations, or participate in discussions on the site. I have changed the names of the Facebook users that figure in Chapter 6 to allow for anonymity.

Despite the extensive amount of other campaign material available, it is the Facebook material that is being focused on in this study. Since the Facebook site enabled a setting where the county council and public met, this material allows for important questions about the specificities of public (girls' and others') involvement in campaign settings, and how this is mediated through digital devices. Moreover, as a lot of the non-digital campaign material was uploaded on Facebook, it was possible for me to study these images in a setting where they were discussed, disputed and transformed. Even though Facebook is the focus of this study, it is worth mentioning that in previous versions, non-Facebook aspects were included in the analytical chapters. The decision to focus solely on Facebook is the result of extensive empirical and analytical work that has sharpened, and limited, the focus of my study.

The second "I love me" campaign consisted of eight textual cancer narratives from young women and relatives, and eight images. The images re-presented the storytellers, and included excerpts from the narratives. The narratives were published online on an "I love me" campaign site.[2] Moreover, during two different campaign periods (May and September

2. When I accessed the county council's web page in April 2016, the campaign was still active.

2013), campaign images depicting excerpts from the cancer narratives, together with images portraying the storytellers, were located on the public transportation system. Additionally, the campaign included a text message service where people could send a text message to obtain a link back to one of the storytellers' narrative. As with the last campaign, the narratives were also posted on the "I love me" Facebook site during the site's existence. However, I will focus on the campaign images and narratives, and not on how they were discussed on Facebook. I have changed the names of the storytellers in the second "I love me" campaign to allow for anonymity.

Combining close reading with a device perspective

In approaching the campaign material, I have combined a close reading (Gallop 2000; Freeman 2010; Lukic and Espinosa 2011; Federico 2016) with an STS device perspective (Law and Ruppert 2013; Ruppert et al. 2013; Savage 2013). Together, this has enabled a material-semiotics approach that looks at the entanglement of visuality, text and materiality (for examples of studies using close reading approaches to visual material, see e.g. Moletsane and Mitchell 2007; Paasonen 2007).

I use close reading to carefully and tentatively attend to the textual and visual material. It is an approach that allows me to engage an ethnographic attitude to this material, and, as part of that, what Celia Roberts (2015: 45) discusses as "ethnographic readings" of textual and visual material. In doing so, I have worked with an approach where I have tried to take seriously unexpected and surprising parts of the material, *as well as* reproductions of the "already known", and of widely circulating politics and assumptions.

As partly already explained in Chapter 2, one way in which I have done this was to attend to marginal parts of the campaign material, and to absences. As feminist STS scholar Jenny Reardon with colleagues (2015) argue, close reading is a helpful method for attending to exclusions, marginalizations and absences in technoscientific texts.

For this endeavor, I have especially found Jane Gallop's (2000) article "The Ethics of Reading: Close Encounters" helpful. In this article, Gallop discusses close reading in a manner I read as in close affinity with an ethnographic attitude. She describes close reading as a method for attending to the surprising and easily overlooked parts of a text. This, she argues, means "giving up the comfort of the familiar, of the already-known for the sake of learning" (Gallop 2000: 11). Concretely, she argues for close reading as a method for attending not only to the main articulations in a text, but *also* to those that would be easy to dismiss as trivial and marginal. It is "a way of learning not to disregard those features of the text that attract our attention, but are not principal ideas" (ibid.: 8).

Instead of reading texts as a matter of looking for what one expects to find or first and foremost for the bigger picture, Gallop emphasizes that close reading allows for an attention to small details. This, she argues, is "the best possible safeguard against projection" (Gallop 2000: 11). In a related vein to Gallop's argument, in *Time Binds: Queer Temporalities, Queer Histories*, Elisabeth Freeman (2010) asserts that reading closely means to look at "the odd detail, the unintelligible or resistant moment" (ibid.: xvi) and "to unfold, slowly" (ibid.: xvii). "To close read is to linger, to dally, to take pleasure in tarrying, and to hold out that these activities can allow us to look both hard and askance at the norm" (ibid.: xvi–xvii).

Like Reardon with colleagues, Gallop and Freeman show, by looking for small details and absences, and by unfolding slowly, close reading enables an attention to details, moments and absences that can disrupt and trouble predominant articulations. Read in this vein, close reading is close to Puig de la Bellacasa's attention to care as an ethico-political commitment to the marginal and easily neglected, and to an ethnographic attitude of being at risk. In line with this, I have used close reading to enable a focus on moments and uncertainties in the material that are easily dismissed as unimportant and trivial. This has meant that I have worked to give up "the comfort of the familiar" (Gallop 2000: 11), to allow the campaign material to surprise me, affect me, and/or teach me about care.

I combine a close reading of the visual and textual components with an attention to what (digital and non-digital) devices *do*. In doing so, I attend to how the app as a learning device, and Facebook devices such as social buttons, facilitate and articulate care. How they, for example, figure as care enablers. By relating to studies on apps and social media (as will be discussed further in Chapters 4 and 6), I have attended to what differences (or not) the digital generates. In doing so, and as touched upon in Chapter 2, I have attended to (coexisting) temporalities of digital media. For example, what happens when a visual is posted as a Facebook status update that girls and others can like, share and comment on? What specific temporalities are generated? In my engagement with Facebook

I attend to how conversations, on the Facebook "I love me" site, were configured and enabled by social button devices, and how this articulated specific matters of care. Accordingly, connected to care, I pay attention to how care comes with specificities when linked to different device-mediated temporalities.

Implicated reading: an affective auto-ethnographic reading

In my engagement with the cancer campaign narratives in Chapter 8, I have performed an *implicated reading* that pays attention to feelings generated by the relationship between the reader and the text (see Pearce 1997; Paasonen 2007, 2010; Liljeström and Paasonen 2010). In close affinity with Gallop and Freeman, Susanna Paasonen (2010: 59) explains this form of reading as a mode of attention that remains "open to surprises and uncertainties while accounting for the affective power, or forces, of the texts studied". This, as she writes, makes it possible to approach texts as a "particular kind of actor with the power to affect its audiences, myself included" (ibid.). Inspired by this, I have performed a reading where I have tried to stay with, and take seriously, my affective responses as an entrance point for doing careful research, and as a resource for potentially valuable analytical insights.

In performing an implicated reading, I have included in Chapter 8 auto-ethnographic writings. Bringing myself into the text is, for me, a feminist methodological strategy for reducing the distance between me, the campaign storytellers, HPV vaccination and cancer. Helpfully, Jenny Sundén (2012) situates her use of auto-ethnography in relation to Haraway's (1997) ethnographic attitude. Doing so, she discusses auto-ethnography as a method of uncertainty and of being at risk. With such an approach, knowing is "shaky, partial, and always in the process of being proved otherwise" (Sundén 2012: 173). Auto-ethnography, formulated in this way, is about holding on to the potential analytical value (and riskiness) of personal, and often affective, experiences. In trying to do so, it is possible for me to attend to how my own affective experiences are "*at stake* in the face of the practices and discourses" (ibid., emphasis added) I encounter, and engage with.

Close to Sundén's argument, I have effected an implicated reading as a way of trying to stay with the risky and partial, yet potentially important, affective responses invoked when I read the campaign cancer narratives.

It is an approach that has made it possible for me to attend closely to experiences and memories that significantly affect how I view, and react toward, the cancer campaign narratives. This has been, as Sundén illuminates, about putting myself "at risk"; to allow myself to get affected, and to take those affective responses seriously. For me this has been about keeping *an openness* to affective responses to enable important analytical insights (see Federico 2016: 98). More details about how I have done this, and more concretely *why* I have found this important empirically and analytically, will be brought up and discussed in Chapter 8.

Interviewing county council professionals: the interviews and interviewees

I interviewed professionals who have worked with the different campaigns. The interviews add an important dimension to the analysis. They enable insights on how people working with the campaigns reflect upon, discuss and articulate their work with the campaigns, and how they reflect upon the campaigns themselves. In this way, the interviews allow for a more multilayered approach, compared with if I had studied only the campaigns.

During my years with this project, people frequently expressed worries or hesitancies toward the fact that I did not interview the girls concerned. In line with such concerns, there are good reasons for doing research that approaches "children as social actors with their own experiences and understandings" (Sparrman 2014: 293). There is a risk of reproducing "adult-centric" research (Hirschfeld 2002). However, I decided to focus on the professionals, as I wanted to enable a study that took the potential nuances and complexities of health communication expert practice seriously, and that approached this with curiosity rather than with predefined convictions. Even if a study including interviews with girls could be valuable, that was not the study I wanted to do. Still, worth noting is that girls "themselves" are involved in the study through analysis of the Facebook site, as girls are one of the "publics" that were active on the site.

I used a snowball selection (see Kvale and Brinkmann 1997). I started with two different people (the health care planner Johan and the communicator Helena) who in one way or another way were responsible for certain parts of the catch-up HPV vaccination in each region. I got in touch with Johan via contacts (a friend of mine knew who he was), and

Helena I found on the county council web page. As it seemed as though both had general responsibility for different parts of the catch-up vaccination, I believed they would be a good starting point. This turned out to be true, as they became important gatekeepers. When I met them, I asked for names of further people involved in the work with the catch-up vaccination or the campaigns. I also used the same method later on in the interviews to get in contact with relevant people. Using this method, it was possible to interview a majority of the people from the county councils who were involved in the work with the catch-up in some way or another.

In the case of Mittland, I have interviewed, to the best of my knowledge, everyone from the county council who were involved in what I define as "the HPV working group". These people have been part of planning meetings. Additionally, the communicator Hanna has, through consultation with the rest of the group, worked with a firm to develop the app's design. She has also been involved in other design matters, such as the development of the advertisements for the app. The information secretary Katarina and the doctor Stefan organized the focus group interviews, and have also been involved in the design process. During this process, they have continuously asked for feedback from the girls in the focus groups about the design of the app. Also the gynecologist Karin was part of the work with the focus groups, this in the form of a medical expert that the girls could pose questions to. She has also proofread factual material used in the app. The health care planner Johan has been involved in planning meetings, and the health care planner Roger has been involved in evaluation work. The school nurse Sara, finally, is re-presented in the app through a video where she figures as a school nurse. Altogether, I interviewed seven people in Mittland.

In the case of Bredland, the organizational structure was different. At this location, there was no working group to interview. Instead, as I have already mentioned, I interviewed the communicator, Helena, who was the one who worked with the "I love me" campaign most extensively (during periods, the "I love me" campaigns took up the majority of her work time as a communicator). She worked with the campaign on a practical level by, for example, organizing the vaccination tour, answering messages on Facebook, and by working with the company designing the campaigns. I also interviewed Klara, Head of Communications, who was responsible for the campaign, and who was one of the main decision makers for the choice of campaign strategies. Additionally, I interviewed a contract administrator and nurse (Linnea) working with the care choice model, who had been involved in planning work for the first "I love me" campaign. Finally, the epidemiologist Emma, who participated in planning meetings for the first "I love me" campaign as a medical expert, was interviewed. She also served as a medical expert in other situations. Altogether, in Bredland I interviewed four people.

Two of the interviewees (Stefan and Katarina) I interviewed together at their request. I interviewed Helena three times about the "I love me" campaigns, since the second "I love me" campaign was launched and designed during the time I conducted interviews, and, later on, a follow-up campaign period was launched. This was a way for me to try to follow through the process with the "I love me" campaigns. As Helena was the person who worked most extensively with the campaigns, it made sense to interview her repeatedly during this period.

A few interviews were short (30–40 minutes), but the majority took longer (1–2,5 hours). The first ten were conducted between January and December 2013. I did one interview in January 2014 as the health care planner Johan was replaced at the end of 2013 by a new health care planner (Roger). In January 2015, I conducted two additional interviews (Emma and Klara), as after going through my interview material again later, I realized that both of them were talked about a lot by my other interviewees. As Emma was an epidemiologist I believed she would add a different perspective than the others (which she did). In turn, as Klara in her position as Head of Communications was one of the initiators to the campaign strategies of the "I love me" campaigns, I believed she would be able to add additional depth to the study (which she did). The full list of interviewees (including their positions and responsibilities) and interviews can be found in my Appendix (names of interviewees have been changed to allow for anonymity).

Interviewing county council professionals has been both intriguing and challenging. Many of my interviewees turned out to be very reflective about their work practices and seemed to enjoy talking about the complexities of HPV vaccination and HPV vaccination campaigns. This can be compared with the few interviews where my interviewees gave short answers and did not expand upon questions in an as reflective a way. In these interviews, I tried using many follow-up questions as a way of approaching the topic from a different angle. This worked in some interviews, but not in all. In some interviews it was hard to even ask follow-up questions since during these moments it felt as though the interviewee "closed" the topic. In such interviews the (in some ways) unequal power relation between my interviewees and me felt very present. However, there was a big difference between the majority of my interviewees who showed interest and enthusiasm for my project, and interviewees who did not and rather left me thinking that they did not understand the possible value of social studies research about HPV vaccination.

Designing and transcribing ethnographic interviews

My interviews have a semi-structured format. I decided on this to enable a conversation between me and my interviewees. I wanted to use pre-constructed themes that could be discussed in the interview setting, but that could also be adjusted based on what was brought up by my interviewees. The semi-structured format was good for this as it enables digression and elaboration from both the perspective of the interviewee and the interviewer (Aspers 2007: 137; Alvesson 2011: 62). In this study, it thus allowed me to attend closely to what happened (what was generated) in the exact interview situation. The semi-structured format allowed me to push the conversation in certain directions, while at the same time remaining open toward what the interviewee emphasized as crucial.

The semi-structured interview method allowed me to take seriously what *my interviewees* cared for through their articulations, and enabled me to learn *from them* about care. This made it possible to approach the interview method as a conduct of what Lotta Björklund Larsen (2010: 65) defines as "ethnographic interviews". Such an approach to the interview method aims for "taking people's stories and accounts seriously" (ibid.: 64). Conducting interviews with an ethnographic attitude is to put myself at risk; to make it possible for the interviewees to challenge my previous convictions.

The interview guide concerned not only the campaigns. It also included questions about the catch-up HPV vaccination at large, and a few more about HPV vaccination outside the setting of the county council. Many of the interview questions asked were concerned with general themes regarding my interviewees' work with HPV vaccination, the process and design of the information campaigns and the work with them. In the majority of the interviews (in all interviews where I thought it was something the person in front of me in some way or another worked with) I also asked questions of how their work was affected by the use of the care choice system – and, thus, how they worked with the care choice model in relation to HPV vaccination and HPV vaccination campaigns.

These questions were meant to make it possible for my interviewees to discuss and reflect upon what they were *doing* at work regarding HPV vaccination and HPV vaccination campaigns. The interview guide also included questions about how they considered and thought about this. Hence, these questions were concerned with my interviewees' *assumptions* and *experiences*. Follow-up questions were often asked to enable interviewees to give examples from their working life or to further develop a theme or a line of thought. It could also be questions requesting clarification of what they meant by something. Moreover, I sometimes used follow-up questions to push my interviewee to elaborate further on some issues. These questions could often be about topics mentioned in passing but that I wanted them to elaborate upon. I sometimes also used follow-up questions to ask them to imagine how county council HPV vaccination work could be different, for example by asking what a campaign focusing on herd immunity in the context of HPV vaccination could look like.

In the interview guide the semi-structured questions were combined with more structured and straightforward questions about the interviewees' position at the county councils as well as the organization of the work around the catch-up at the county councils. These questions were often followed up by questions of a less straightforward kind since my interviewees often expanded upon such details into more elaborated and narrated ways. I ended the interviews by asking more general questions about HPV vaccination "at large". I adjusted these questions based on happenings and debates taking place in society at that time. For example, when I interviewed the communicator Helena and the school nurse Sara, a highly critical Swedish documentary on HPV vaccination had just been screened. To enable them to discuss HPV vaccination in an "up to date" manner I asked whether they had seen it and if so, what they thought about it.

All interviews have been recorded and transcribed word for word. When transcribing I have also put in brackets silences, laughter, hesitations, surrounding noise as well as how materialities (such as the campaign images and the HPV app) were mobilized in the interviews (for example by writing things such as "Stefan opens his email on his laptop …"). This has proven worthwhile, as these small details have often, as I will discuss as part of the next section, turned out to be important empirical findings. Moreover, after the interviews, I wrote short memory memos about the interviews. These concerned things like my encounter with interviewees, the location for the interview, and how I experienced the interview.

The interview setting as agential: attending to subtleties and materialities in interviews

In February 2013, I met the school nurse Sara in her office at the high school where she worked to talk with her about her involvement in the work with the HPV app. During the interview I asked her why she thought the HPV app was needed. She replied:

> The idea is interesting but I can't say that this will be the big thing for the future [...] But perhaps I'm not allowed to say that here [laughter] [...] But perhaps that isn't what you want to hear.[3]

By laughing and saying that perhaps I do not want to hear her saying that the app is not completely fitting, Sara answer provides insights concerning the interview method. Being asked about why the app is suitable, she was led by my question to make excuses for feeling that she could not answer it in "the right way". Because of how I asked the question, Sara felt like she could not give me an answer, something that made her laugh. Based on my assumption that people in the HPV app group would consider the app as necessary, I posed a question that relied on that very assumption. This in turn made her express her uncertainty about the app. The answer was generated by the very interview situation: the question posed by me, and Sara's expectations of what I wanted to hear.

This short excerpt from the interview with Sara serves as a good example of how what my interviewees said in the interviews is affected by the interview setting, including by myself. Sara's trouble with answering the question illuminates how her answer was not "premade, ready to share with the interviewer" (Müller and Kenney 2014: 554).

Helpfully, Linnea Bodén asserts that in interviews, "the phenomenon will always be produced in the relations between data, method/ methodologies, research questions and theories" (Bodén 2015: 195). In line with this, I am interested in paying attention to "the subtle effects of the interviews that are often ignored or set aside as the interview data

3. This quote is also used in Chapter 5 but is there for partly different reasons.

are transformed into research findings" (Müller and Kenney 2014: 539), which means that "[t]he question asked, the time, pace, and space of the interview; the voices heard or unheard; the actions recognized or ignored – will enable some intraactions[4] and impede others" (Bodén 2015: 195). Turning back to the conversation between Sara and me, the subtle disruption of her excusing laughter, instead of being a problem, may serve as a moment that says something about the phenomenon studied. Much in line with a close reading method, I have tried to pay close attention to moments and parts of the interviews that are easily overlooked, and that might disrupt anticipated answers and predominant articulations. This is especially evident in Chapter 5, but also in Chapter 9.

The interview situation is often considered to be interpersonal in its format: it is the meeting between me and my interviewee that matters. This is also what I have emphasized so far through the example from my interview with Sara. The idea of interviews as first and foremost interpersonal can, however, be questioned. Bodén (2015) does so by bringing materiality into the interview setting. She argues for the concept of *intraviewing* as a concept for approaching the interview setting as an "intra-active event" in which diverse agents/actors are enacted, and that are part of what enacts, the interview setting. *Intra-active* is a word she borrows from feminist scholar Barad (2007) to talk about the entanglement (and becoming) of the material and the discursive.

In my interviews I did not explicitly try to do "intraviews", but there were several moments where materialities were made present in ways that, for me, has turned out to be worth unpacking further. Even though specific examples will be brought up in the analytical chapters, for now it is important to emphasize that such moments can shed light on how researchers and interviewees *together with* materialities bring about, and enable, specificities in interviews. Concretely, an important focus for this study has become on how materialities mattered in these situations, and how this enabled matters of care. Following the study's theoretical approach, such materialities will be discussed in terms of *devices*. How material devices oriented, and transformed, the directions of our conversations will be brought up further, explained and discussed in Chapters 7 and 9.

4. This is a notion from Barad (2007) used to talk about the inseparability of materiality and discourse. Bodén writes *intraaction* without a hyphen and not *intra-action* as it is usually spelled.

Turning diverse empirical materials into analytical themes

To be able to navigate through my empirical material I have used coding. Coding can be understood as a way of organizing difference and sameness in the material (Aspers 2007). Through coding, I could see connections between different parts of the material, connections that later were turned into analytical themes for the empirical chapters. To be able to perform coding I used the computer software NVivo. This program coded my interview material, parts of the Facebook material, the cancer narratives and the campaign posters and pamphlets. I made two (as it is called in NVivo) different coding cases: one for each county council. This was not a given decision as I did not know initially whether it was best to write separate chapters for the different county councils or not. As it turned out, I realized that it worked best to write two chapters for each campaign (that is, two chapters about the app, two about the first "I love me" campaign, and two about the second "I love me" campaign). In hindsight, I can thus see that it was a good idea to make two separate coding cases in NVivo.

To get an overview of the material, before the first round of coding I read through the material carefully, and made notes. During an initial round of coding of the interview material, I did an initial coding where I used themes from the empirical material as names for the codes. During a second round, several codes were merged and separated as I could see that they belonged with each other. During this second round, several codes received more "abstract" (and often theoretically connected) names. This was the case when a more abstract and/or theoretically grounded name worked well as a signifier for the new collection of merged empirical material. Several codes, however, still had empirically close names. For example, the code *trust* was developed based on my empirics (it is a word some of my interviewees used extensively) but was, as a name, also connected to previous research on vaccinations. The code *being where girls are*, in turn, was fully based on my interviewees' formulations. All codes were modified, changed and regrouped during the period of coding as new interview material made it possible to see how, on the one hand, things that I previously thought could be under the same code needed to be in separate ones and, on the other hand, how other codes could be combined into one.

When coding the cancer storytelling texts, several codes got empirically close codes such as *breaking down* and *future* but also more theoretically grounded codes such as *affective relations* and *anticipating risk*. As with the interviews, I worked with two rounds of coding where names were changed and material was merged and separated.

In the process of coding the visual material, I was guided by STS scholar Adele Clarke's guidelines for how to perform coding in connection with visual material. Following Clarke's guidelines, I wrote *analytical memos* by means of which narratives about each visual (e.g. campaign images) could be produced. This required me to put into words what I saw when analyzing the campaign material, and made it possible to make this material more like other material while at the same time allowing for its specificities (Clarke 2005: 224–226). It also made it possible to treat the images as a matter of close reading, as this allowed me to write down, and take seriously, details in the visuals. Guided by Clarke I did two types of memos: *big picture memos* that enabled me to describe the visuals fully; and *specification memos* that made it possible to "break the frame" of the visual material so that I could see the images in multiple ways and address any absences and exclusions. The specification memos were written on the basis of questions suggested by Clarke, such as how the subject of the visual is framed, who the intended and unintended audiences of the visual are, what possible absences I could think about and if there are any remediations where digital technologies are connected to "older" media (Clarke 2005: 227–228).

I have tried to perform coding that enables an analysis that addresses both content and materiality. For example, regarding the app, a code such as *sex as risk* has to do with content and *push notification* has to do with material capacities. The coding of the app material was both empirically grounded and connected to previous social science research on apps. The Facebook material is, as already noted, *extensive* and I have not coded it all (but I have read through it several times and made notes). Regarding the Facebook material, codes such as *responding with facts*, *vaccination critique*, *vaccination fear* are not Facebook specific codes. In contrast, *numbers as important* and *encouragements to share* signify acts of liking, sharing and commenting on Facebook (and hence, material device capacities). In practice, however, since, for instance, vaccination critique and vaccination fear were mediated via Facebook social buttons, these codes also point toward Facebook specificities. By separating them as different codes, however, I was able to pay close attention to both content and materiality without risking neglecting one of the sides. This separation served as a pragmatic solution. When analyzing the material, I have "re-merged" content and materiality.

I have read all my material slowly and repeatedly as a way of trying to avoid simplified and hasty analyses and conclusions. I have also, after having coded the material, repeatedly gone back and read the entire transcripts. This has allowed for an analytical approach that remains open to surprises, nuances, specificities and differences. With the Facebook material, I have repeatedly gone through all the screenshots. When starting major revisions of older versions of an analytical chapter, I have begun the process for that chapter with reading the whole of the relevant material. After having done so, I have gone back to the codes. Sometimes this has meant that I have recoded bits of the material. This has made it possible for me to, based on the progress of the analytical work, find new parts of, and connections to, the material that during previous processes of coding I overlooked or dismissed as unimportant. During the process of analytical work I have thus continuously aimed to "undo, trouble, keep open and push boundaries" (Latimer and Skeggs 2011: 401) – including the boundaries and closures through which I had previously divided my material, and analysis. At the same time, the coding process does, per definition, define and pattern empirical material. It *is* a matter of "closing down" things. What I have tried to do is to keep questioning *how* I have done so.

As explained in Chapter 1, the empirical chapters are divided into a set of three "empirical parts" which are organized based on the three campaigns. Each part centers on one campaign, and consists of one "campaign chapter" and one "interview chapter". This division is the result of extensive analytical work and rework. During the years with the project more thematic structures have been tried out (e.g. one chapter about sexual dimensions, one about care, and one about trust). Until the final stage with the project also the campaign material and the interviews were combined in the same chapters. I decided to change the study's structure due to several reasons. The different campaigns are different from each other, and these differences became clearer when I was allowed to pay close attention to the campaigns separately. Moreover, to separate the interview material from the campaign material enabled me to attend carefully to the specificities of the different kinds of empirical material; what they allowed me to see, say and do.

As with the structure of chapters, the thematic focus of the chapters is the result of extensive analytical work. The topics of the different interview chapters are chosen based on what were the central matters in the interviews. For example, in relation to the HPV app interviews, my interviewees' main focus was on a need to communicate factual information to girls. Therefore, this became the main theme of that chapter. In connection to the first "I love me" campaign my interviewees' narratives were centered on how girl empowerment can be a way of reaching girls, and therefore this became the main focus of that chapter. In line with my analytical and theoretical focus on marginal, absent and alternative matters of care, I also focused on moments of frictions in, or divergences from, these main narratives.

Discussion: methods' risky and promising agencies

In concluding this chapter, I want to relate what I have brought up so far to a broader discussion on methods in STS. At the beginning of this PhD project I wanted to do participant observations at Bredland and Mittland County Councils to be able to catch the details of everyday work with the catch-up HPV vaccination. Back then, it felt like that to catch the complexity of HPV vaccination *in practice*, I needed to be out there and see what was *really* going on. I had no access problem, as my interviewees were willing to let me do observations. But, as it turned out, almost nothing was happening at all on the HPV vaccination front at the county councils. One idea I had was to go along with the vaccination trailer team when they were going out to schools, but after a couple of months it was decided that a second tour with the trailer would be too expensive and it was cancelled. This was just one of several ideas I had that turned out not to be feasible. After a couple of months, I had to give up and decided to focus fully on other methods.

Why did I want so strongly to do observations? I realized that this did not merely have to do with me wanting to catch HPV vaccination "on the ground". It had also to do with a lingering fear of my work turning out to be less valuable – less good – if I only conducted interviews and studied the campaign material. In trying to unpack what this fear might signify, STS scholar Sonja Jerak-Zuiderent (2014) helpfully talks about a methods hierarchy in STS between observations and interviews:

> Although [participatory observation] has proven fruitful, it poses its own challenges […] [T]his route easily leads to privileging ethnographic observations as more real and complex than accounts generated by other methods […] Privileging of ethnographic observations points to a lingering simplistic realism in STS, an implicit empiricism that has been critiqued in relation to other scientific practices as a "god trick" […] [I]nserting the epistemic hierarchy of ethnographic practice into interview stories denies their "reality". (Jerak-Zuiderent 2014: 904.)

When STS researchers attend to complexity and multiplicity, observations are indeed a common route. As Jerak-Zuiderent (2014) points out, the risks with this are not only that observations are privileged as more real; it also risks positioning the researcher as the one speaking the truth about what is really going on *out there* (that is, that the researcher performs the god trick). Jerak-Zuiderent asserts that this *denies the interviews' reality.* In a related vein, feminist STS scholars Ruth Müller and Martha Kenney (2014: 541) use interviews as the basis for a call for a need for "keener sensitivities to the effects of our methods as a way to orient and re-orient our research projects". Following this, they stress the importance of "taking seriously the risky and promising agencies of our research apparatuses" (ibid.: 553).

A call for keener sensitivities toward the risks and promises with specific methods is also how I interpret Cartwright (2014: 254) when she asserts that it is a problem that STS first and foremost takes into consideration re-presentations when "they are entered into an analytic framework in which they provisionally cleave to other matter". As with interviews, Cartwright illuminates that there is a tendency in STS to attend only to re-presentations as being enacted in *other* practices (often studied through observations). Here, re-presentations tend to come to matter when they move around, are translated.

If different methods entail specific promises, troubles and risks – if each method entails "a wonderfully detailed, active, partial way of organizing worlds" (Haraway 1997: 90) – it is crucial to foster keener sensitivities toward the implications of this. I believe that by addressing this further, it is possible to learn more about how different methods, in different ways and in different situations, take part in making worlds. Following such commitment, in this study I try to hold on to the possibilities and limitations of the diverse materials and methods I work with. That is, I try to take *their* realities/worlds seriously, and what these allow me to see, say and do.

EMPIRICAL PART I
REACHING THE GIRLS

Health promoters envision smartphone apps as promising devices for reaching teenagers "where they are" (Lefebvre 2009; Levine 2011; Ralph et al. 2011) with "accurate" and "good" HPV vaccination information (Hill et al. 2013; Zimet et al. 2013). The HPV app is an example of such a device used to provide information to girls. Empirical Part I focuses on this app, and in doing so it discusses relations between health information, health communication devices, girls and care. It discusses *how* girls are imagined to be reached by the app, and *what* HPV vaccination related information they need to be reached with.

The app was designed by Mittland County Council in Sweden in 2012, but is today promoted by, and used in, several other county councils. The app is a result of focus group interviews that Mittland County Council did with girls in high school. In the focus groups, the county council asked questions concerning how the girls thought young girls wanted to get information about HPV vaccination, and how girls can be reached today. The very idea of designing an app was generated from the girls in the focus groups.

The app consists of six different parts. The first part is a short movie featuring a school nurse telling girls why they should get vaccinated. The second part provides information regarding why it is good to get vaccinated. The third part consists of a "did you know that …?" list that gives information about cervical cancer, HPV and HPV vaccination. The fourth part provides information about how to get vaccinated, hence answering the question "how do I do it?" It also includes a calendar in which girls can insert when they received the first vaccination shot. If they do so, they will receive a reminder when it is time to take the next. The fifth part is a quiz where, by answering "yes" or "no" to different assumptions about HPV, HPV vaccines, HPV vaccination, sex and cervical cancer, users can learn about the vaccination. The sixth and final part provides the opportunity to send feedback to the county council. The app is designed with a background that is partly black and partly comprises orange hearts, with flowers in a hand-painted style and with the text "LOVE" in a similarly hand-painted style. It includes two people that look as though they are about to kiss. In addition to orange and black, the app has some text in white (see Figures 1, 2 and 3).

Empirical Part I consists of two analytical chapters: "The HPV App and a Care for Neglected Things" (Chapter 4) and "Facts, Fears and Frictions" (Chapter 5). Whereas Chapter 4 analyzes the content and the materiality of the app, Chapter 5 draws upon interviews with the HPV app working group. In the chapters I approach the HPV app as a care enabler that entails a promise of facilitating girls' capability of taking care of themselves through making a vaccination decision. In different ways, the chapters complicate such an idea by discussing *for whom* the app enables care, and *what* forms of care it enables.

In Chapter 4, I relate the HPV app to other social studies and humanities research on apps, and ask why apps like the HPV app that *provide* information rather than *generate* user data seem easy to overlook and neglect. In contrasting the HPV app to "hyped" data generating apps, I discuss the HPV as inhabiting coexisting temporalities of mediation and care.

In Chapter 5, I focus on a predominant matter of care that appeared in the interviews, which articulates that vaccination fears need to be counteracted with accurate and impartial information. By making present marginal articulations, which trouble such a "clear-cut" vision of fears versus facts, I discuss the ethico-political potential of alternative matters of care "*within* otherwise dominant configurations" (Martin et al. 2015: 634, emphasis added). In doing so, I enter a discussion concerning how a slowing down of the plot can trouble an idea that actions need to be taken *now* to counteract fears.

Altogether, Empirical Part I centers around the HPV app to discuss different links between girls, temporalities, feelings and care. The two chapters unpack and complicate the HPV app's status as a care enabler,

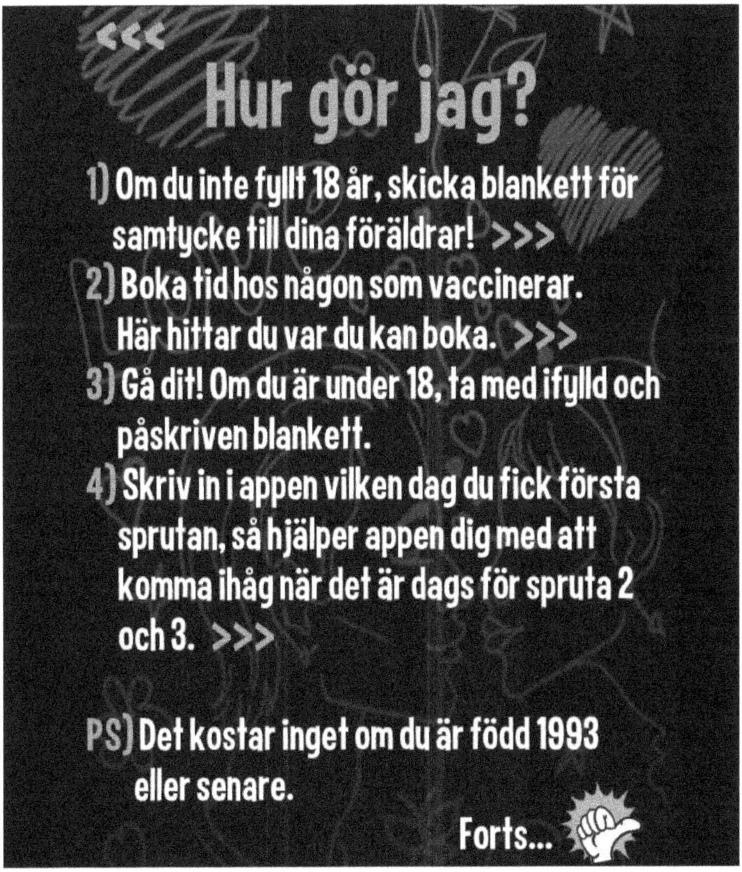

Figures 1–3. Screenshots of the HPV app.

and make present alternatives *already circulating* within the empirical material. In focusing on temporalities of care, the chapters set the basis for the discussions in Empirical Parts II and III.

4. The HPV App and a Care for Neglected Things

"Did you know that … HPV is very common and that most sexually active adults have had it[?]", the HPV app asks me. Opening my web browser, on the site *HPVkoll.se*,[1] I get related information through a similar "did you know that …?" list. "Did you know that … about 40,000 women in Sweden are afflicted by cervical lesions every year?", *HPVkoll.se* asks me.

This chapter discusses how the HPV app shares similarities with other health campaign, and health information, devices. As the *HPVkoll.se* site exemplifies, such devices can be regular web pages, but they can also be, for instance, posters, pamphlets, questionnaires, paper documents and reminder letters. As illustrated with the "did you know that …?" list, the HPV app primary *provides* HPV vaccination information to the girls concerned. As a health information device, it is designed to enable a vaccination decision.

This makes the HPV app different from the apps most often discussed in other studies. The apps often considered *generate* personalized user data, data that feeds into a seemingly ever expanding production of "big data" (Jethani 2014; Millington 2014; Adams and Niezen forthcoming). In the context of health and illness, self-tracking apps especially have gained extensive interest (see e.g. Lupton 2013, 2014; Millington 2014; Till 2014; Lupton and Jutel 2015; Lupton and Thomas 2015; Maturo and Setiffi forthcoming). These apps are often discussed as new health communication devices, and are contrasted with old ones, such as posters and pamphlets (see e.g. Johnson 2014). Different from these apps, the HPV app does not track its users' actions. Moreover, its data is not big in volume or in terms of velocity, and does not feed into bigger flows of data

1. "HPVkoll" translates from Swedish to "HPV informed".

used by third parties.[2] When downloaded, most of the HPV app's parts do not even require Internet access.

In the article "Apps as Artefacts", Lupton (2014: 608) highlights that many of the apps she found during her research were "very simple". They were "providing information on a specific medical condition or treatment" (ibid.) without including self-tracking and data-generating components. As Lupton touches upon, there exist tons of health and illness apps that are perhaps used once or twice and that do not track users' bodily movements and actions. Through a quick look at *App Store* and *Google Play*, I found a range of apps that, similarly to the HPV app, first and foremost *provide* information, such as apps for looking up symptoms and for providing information about specific diagnoses or "health risks". For instance, the *HPVsearch* app includes bullet point lists partly similar, as I will discuss, to the design of the HPV app. To take another example, the app *Health & Illness* "gives you tons of useful information on health and illness" by providing short excerpts of information related to different health and illness topics without enabling the user to add personal data.

Even if Lupton herself does not further address apps like these (she is more interested in the big data and self-tracking dimensions of apps), the small snippet in her article mentioning the range of "very simple" apps opens up for consideration how it is that these apps can be so many, and yet are so easily brushed over and left unanalyzed. I believe this provides a reason for staying with the trouble of apps.

Speaking implicitly to such concerns, in the recent review article "Quantifying the Body and Health: Adding to the Buzz?", STS scholar Michael Penkler (2015) expresses caution against a current STS rush toward self-tracking and big data. He stresses that increased sensitivity is needed toward *how* engagements with these "new" technologies risk reproducing a technoscientific "hype". In doing so, he also points toward a potential risk of allowing an STS narrative about self-tracking and

2. It is not a given as to how one should define big data. For the sake of this chapter, I follow the definition put forward by STS scholar Evelyn Ruppert with colleagues (2015). They write: "While variously defined, Big Data generally refers to digital content stored in social, commercial, scientific, and governmental databases and often generated as a by-product of digital transactions, communications, interactions, and so on. According to the most popularly referenced definition, what makes this data distinctive is not only its volume but its velocity of generation (the speed of collecting data in 'real time') and variety of data sources and formats (increasing array of data types from audio, video, and image data, and the mixing and linking of information collected from diverse sources)." (Ruppert et al. 2015: 1.)

big data to include an assumption about a teleological or epochal shift toward "the new". However, while Penker argues for a need for critical attention to *how* STS scholars engage with self-tracking devices (such as self-tracking apps), I assert that it is also important to attend to *other* digital devices that seem easy to neglect or overlook. It is, I argue, important to take these devices seriously in their own right, as they might enable insights that can trouble technoscientific "hypes".

Following this, as a case of what Lupton (2014: 608) defines as a "very simple" information app, I argue that the HPV app needs some attention. More preciously, by drawing upon Puig de la Bellacasa (2011) who focuses on an ethico-political commitment to generate care for undervalued and neglected things, I assert that the HPV app needs some *care*. Therefore, I will explore what is enabled by a care for the HPV app.

As I do in all chapters, I will address care on several dimensions. In addition to "my" care for the app as an example of a neglected type of app, I focus on how care is already circulating in, and through, the app. As I will show, the app can fruitfully be understood as inhabiting a vision of being a care enabler. Through its data/information, the HPV app is articulated as facilitating girls' capacity to get vaccinated, and to care for themselves. Yet another dimension of care is how the content of the app includes re-presentations of care.

As the app is a device that *provides* information, an analysis of that very information is needed. I will here combine attention to the material capacities of the app with a close reading of the app's content, including its visual aspects. This enables an analysis of *what* information is mediated and *how* it is mediated. Thus, I provide a detailed analysis of the different parts of the app. I discuss the video in the app, the "why getting vaccinated?" information, the "how do I do?" information, the "did you know that …?" assumptions in the app, its quiz, and its feedback function. I end the chapter by zooming out from the detailed focus on the content and material capacities of the app, and focus on what the chapter says about a discussion of care for neglected things, and this in relation to coexisting temporalities of mediation – and of care.

A first meeting with the app: to the movies

The first thing that happens when you open the HPV app for the first time is that it tells you that it would like to send you "push notifications". The notifications (three in total) only include info about when to get the next vaccination shot. I will discuss the push notifications later.

The second thing that happens when using the app is that a short movie (26 seconds) starts playing. The movie features the school nurse Sara telling two girls why HPV vaccination is something good. This only happens the first time the app is used. During further use, the user needs to click on the video to make it start.

The movie starts with the school nurse performing typical school-nurse tasks: she takes pupils' temperature, checks blood pressure and prepares for an injection. She is re-presented as a school nurse through the visual presence of different medical devices: the sphygmomanometer, the syringe, the clinical thermometer, and the eye chart. Through these practical nursing tasks and devices, she is shown doing *care work*. This inhabits a gendered idea of nursing as a caring profession. That is, it includes "gendered norms of care labour" (Murphy 2015: 731).

In the beginning of the movie, the arms and hands of a teenage girl are visible. The girl is wearing festival wristbands, something that serves to signify teenage life. Gender is done: her nails are polished and she wears typically femininely coded earrings (also a whiteness norm is performed as all subjects in the video are white). Up-beat music is playing in the background, something that further signifies teenage life.

In the next scene, the school nurse asks two girls whether they have vaccinated themselves against HPV. As was the case with the girl wearing festival wristbands, the girls are visualized according to predominant gendered features when it comes to hair, make-up and clothes. This serves to invoke them as "typical" girls who are representative of the target population of girls. On the question of whether they have gotten vaccinated, the first girl answers yes, the second says no. In response to this, the school nurse turns her attention from the girls and looks into the camera – and, in doing so, moves from the girls in the movie to the whole imagined population of concerned girls – and says: "Have *you* thought about doing it? It gives really good protection against cervical cancer".

When moving her attention from the girls in the movie to girls "out there" ("have *you* thought about doing it?"), the nurse shifts her care from being a care for the two girls in the movie (a care for individuals), to a care for the collective (population) of targeted girls "out there". By moving her focus from the girls in the movie, the audience, and therefore the targeted population, is addressed. That is, "you" becomes equalized with the imagined population of targeted girls. The movie presents "you" as the centered decision maker, but "you" can be any one in the imagined population of targeted girls. Thus, the re-presentation of the girls and the school nurse in the video is made possible through a connection to other subjects "out there", and the movie *depends on* an *absent present* population of imagined targeted girls; they are addressed without being visually present.

Movies have long been used as health campaign devices in Sweden (Thorsén 2013), and elsewhere (Tulloch and Lupton 1997; Ostherr 2013). The movie as an "older" health communication device is part of the "newer" HPV app. The movie *enables* the HPV app, as these media devices *coexist* in the app. This troubles any clear-cut demarcation for what is "new" and "old" about the HPV app. The movie medium is represented in the app, and *the HPV app is an app, but it is also a movie.*

"Why get vaccinated?": a digital information pamphlet

When the video has stopped playing, the user can decide what to do next. More concretely, the user can decide between clicking different parts, such as: "why get vaccinated?", "did you know that …?", "how do I do?", "quiz", "help us!" or playing the movie one more time.

Under "why get vaccinated?" the user can read about different reasons for getting vaccinated. The information is mediated through a "questions and answers" design. The information is given as a list, and all the text is provided on the same page. Its design and content is very close to what a health information pamphlet often looks like. "Question and answer" designs are common in the context of health campaigns (see e.g. Dugdale 1999: 128). Thus, the app can be understood as a *digital information pamphlet. It is an app, but it is also a pamphlet.* These media devices coexist in the app. The app *refashions* the information pamphlet.

This part of the app tells the user that, for example, it is good to get vaccinated as "it prevents cervical cancer", and that HPV vaccination "is safe", as "over 30 million people all over the world have gotten vaccinated, thus the vaccine is reliable".

Stating that "it's safe" depends on the articulation of a *past* that has accumulated 30 million vaccinated people. Through this formulation, a past is invoked and brought to life. Moreover, by stating that the vaccination prevents cervical cancer, it is also about *anticipating* future cancer. Thus, a specific future is also brought into the picture. That is, through the information, the app *folds* past, present and future time.

In bringing up that 30 million *people* have gotten vaccinated, the information in the app makes it possible to imagine that "people" are not all girls or women. As this is a different reading than being in line with a girl-centered one, it expands the vaccination into a matter of care that potentially includes male subjects. As such it differs from, for example, the content of the highly gendered direct-to-consumer (DTC) ads for Gardasil in Sweden where girls were encouraged to join a "a worldwide collective of women and adolescent girls that together are 'fighting against cervical cancer'" (Lindén 2013b: 88). In the app, "people" opens up an alternative narrative to a female-centered one.

The "why get vaccinated?" section is focused on cancer, and not on HPV and its sexual dimensions. The app communicates: "It can feel inconvenient to have to go to the vaccination clinic three times, but to prevent cancer, it's worth it!" Stating that it "prevents cancer" makes absent how HPV vaccines are *estimated* to prevent only 70 percent of cervical cancer cases each year. Moreover, in the name of cancer, the vaccination is emphasized as important and, once again, "worth it!" (also implicating that *you* are worth it!) Additionally, in focusing on cancer, HPV and sexual activity are made absent and a common "cancer frame" is drawn upon. However, this needs to be read in relation to other parts of the app. As I will discuss, there is, in fact, a quite extensive focus on HPV and sexual activity in the app.

The "why get vaccinated?" section continues with the words that "since the vaccine is so good, girls get vaccinated in the fifth–sixth grade in school" but "also you [that is, born between 1993 and 1998] of course need the protection!" Together with the formulation mentioned earlier that "it's safe", stating that the vaccination is "so good", and then following it with saying that "also you" need the protection; HPV vaccination is invoked as a given decision. By making absent references to, for example, possible side effects, the decision is presented as a no brainer: it is pure good. The focus on *you* stages the app user (most likely, a girl) as the decision maker, but the decision is not really re-presented as a difficult one.

Still under the "why get vaccinated?" section, the formulation "we help you!" as "we remind you via the app when it's time for your next shot and help you with the consent form your parents are to sign" comes next. This section ends with the words "take care of yourself and good luck!" In emphasizing that "we help you!" the vaccination decision process is re-presented as involving the county council and the app. The focus is on how the county council *through* the app helps with the decision. In ending with a focus on self-care ("take care of yourself!") a shared care responsibility between the county council, the girls and the app is articulated. In contrast to the focus on *you* before, the decision is now articulated as shared. This shared responsibility depends on – as I will move on to discuss – a delegation of care responsibility to the app's calendar and push notifications. The articulation of a shared responsibility differs from a predominant focus on girls' responsibility to get vaccinated against HPV.

By providing necessary information about why it is important to get vaccinated, the app (and the county council) are envisioned as *enabling* girls to care for themselves through vaccination. As such, the app articulates a promise of being a care enabler that facilitates girls' ability to care for themselves. Importantly, this promise of the HPV app as a care enabler includes promotion of a specific form of care: HPV vaccination as girl-centered self-care.

Proposing a "to-do" list: a paper document and some push notifications

Under "how do I do it?" the user receives four different numerically sequenced bits of information about how to go about getting vaccinated (Figure 3, page 83). These include links to the vaccination consent form (which parents need to sign if the teenage girl is under 18), and to a website linking to the different county councils' vaccinators. Through presenting the vaccination process as four straightforward steps, the app articulates it as something easy and logical. Just follow the "to-do" list, and the app promises that things will be fine.

Whereas the app in general is focused on teenage girls, the link to the vaccination consent form device provides a moment of parental presence. It highlights a tension between a currently prevalent, and in the app evident, welfare idea of the importance of enabling children to make decisions about their own lives (e.g. Sandin and Halldén 2003), and the legal regulation stating that parents need to be the ones that *actually* decide. However, by presenting the process as a clearly sequenced and easy one, the fact that parents need to sign the form is re-presented as an easy step on the way.

The consent form is a paper document. If the user clicks on the app's link to it, a web page is reached where one can download, and print, the document. This paper document is one of the parts of the app; it *enables* the app. That is, *the app is an app, but it is also a paper document.* Under the same section, the user is encouraged to "write in the app what day you got your first shot" as "the app helps you remember when it's time for shots 2 and 3". This is followed by a link to the app's calendar. This part of the app can only be used if the user is online. Clicking on the link, the user is encouraged to "report when you got your first shot" to the app's calendar. If filled in, the calendar gets translated into the push notifications that I have already mentioned.

The push notifications are moments where the app makes itself present to the user. "Unannounced" it calls for the user's attention. By sending push notifications as pop-up reminders (and in case of the first one, as a question about permission to send push notifications in the future), girls are reminded that they ought to get vaccinated. Through the push notifications, girl-centered vaccination responsibility is repeatedly *confirmed.*

Following this, it can be argued that the push notifications articulate what Sophia Alice Johnson (2014: 333) defines as "push responsibilisation". She explains the concept as signifying an individual health responsibility facilitated by daily or weekly updates sent by an app. These updates, she writes, encourage users to follow specific health information, and often add new bodily data. The push responsibilization performed in the HPV app differs a little from this. It does not facilitate long-term use (no daily or weekly updates), and the only personalized data it includes is the information about when the user received her first shot.

The push notifications in the app include a delegation of care responsibility *to the app*. When asking the user to allow it to send out push notifications, it also asks it if it is acceptable that it *takes care of* remembering when it is time for the next shot. By delegating the app to remember *for the user*, the app works as a memory device in a manner reminiscent of how José Van Dijck (2008: 120) discusses how "media technologies play a constitutive role in the production of memories". The push notifications include not only reminders about when to take the shot, they also simultaneously take part in *constituting* the user's memory. In taking care of remembering for the girls, the app works as a care enabler that facilitates girls' ability to perform this specific form of care: self-care *as* HPV vaccination.

Care responsibility is not only delegated *to* the app. The girls are the ones that actually need to decide to take the next shot, and who need to go to a vaccinator. The app delegates a care responsibility *back to* girls.

This delegation from the app to the girls occurs when a push responsibilization is articulated.

Sending out reminders (as the push notifications do) that health interventions are provided is not something new. On the contrary, this is common in health care in general, including vaccination interventions (Leask 2002). The most common way is still perhaps to send appointment notifications and reminders via regular post. These often articulate a similarly "simple" message as do the push notifications: you have a possible doctor's appointment waiting. That is, the push notification reminders and "regular" reminder notifications are similar, and the push notifications are not simply "new". Reminder notifications are *remediated* digitally. *The HPV app is an app, but it is also a reminder notification device.*

To sum up, the "how do I do?" section of the app provides a "to-do" list that consists of a mixture of different devices. It includes a consent form (mediated through a link to a regular web page), three push notifications and a calendar. These devices coexist in the app, and are what assembles the HPV app into a care enabler. The care the app enables is dependent *on* coexisting media devices, which include a mixture of "newer" media, and "older" media devices. *The HPV app is an app, but it is also a paper document, a reminder notification device and a calendar.*

"Did you know that …?": an HPV vaccination encyclopedia

The HPV app provides factual information about HPV, cervical cancer and HPV vaccination. This is especially the case under the section "did you know that …?" in the app. Here, different HPV vaccination related statements are listed. It is, for example, stated that "there is good knowledge about the HPV vaccine's safety" and that "more than 89 million dosages have been given". This articulates the vaccination as safe. Adding the info that "[a]bout 450 women in Sweden are inflicted every year by cervical cancer" Gardasil is re-presented as a safe option that prevents you from becoming one of those 450 women. The information is re-presented as neutral and objective facts "from nowhere" through the emphasis on numbers of women and dosages, and through articulations such as that the vaccines *are* safe.

Under the section "about cervical cancer" it is stated: "The development of cervical cancer often takes several years. Dangerous lesions that can cause the disease are most often discovered in time if you go to regular Pap smears". Standing in sharp contrast to the no-brainer decision formulation discussed earlier, through this wording the Pap smear is re-presented as most often preventing cervical cancer (as it most often discovers dangerous lesions in time). "In time" here refers to before they get cancerous (and implies a quite ambiguous temporality – "in time" is hard to fully anticipate). It is articulated that the Pap smear is *most often* enough. This enables a message stating that the Pap smear is good, it is just not *good enough* to *always* do all the work itself.

The information under the "did you know that ...?" section of the app relates HPV vaccination to sexual activity. It is stated that "getting inflicted by HPV during sex is the most common cervical cancer cause". Moreover, it is stressed that HPV is related to being "sexually active". In fact, under the headline "about HPV", HPV's relation to cervical cancer is not the focus at all. Instead it is mentioned how common is HPV, that it can generate genital warts, how and when HPV was discovered and, once again, that it is sexually transmitted.

The focus on HPV and sexual activity stands in contrast to the predominant focus on cervical cancer (and absence of sexual dimensions) in other campaigns. For example, and similarly to campaigns from, for example, the US (Braun and Phoun 2010) and Canada (Connell and Hunt 2010), the first "I love me" campaign in Sweden re-presented the HPV vaccination as about cervical cancer and not sexual activity (see Chapter 6). The app's focus on HPV as sexually transmitted differs from this. Instead of being absent, sex is here present as a matter of sexual infections.

As is the case with the most parts of the HPV app, under the "did you know that ...?" section, all information is given up-front in a list. The emphasis is on providing "simple" and "straightforward" information about HPV vaccination. As I mentioned in the introduction to this chapter, "did you know that ...?" lists are commonly used as health information devices. More concretely, the "did you know that ...?" format is often used in non-data-generating apps (such as *HPVsearch* and *Health & Illness*, as already mentioned), in patient web pages listing info about diagnoses, symptoms and treatments, and in health information pamphlets. This illustrates a consistency between the HPV app and "older" forms of health communication. What is more, the app is not only reminiscent of these other forms; it is *enabled* by the temporal consistencies between different media devices. As already partly discussed in relation to the "why get vaccinated?" section, *the HPV app is an app, but it is also a patient pamphlet, and a web page.*

Get the facts right: a quiz

The quiz in the app consists of "common assumptions" about HPV vaccination, HPV vaccines, sexual protection, and cervical cancer. Playing the quiz, you as an user get five different assumptions. Every time you play, the app randomizes between the whole range of assumptions existing in the app. The range of assumptions includes for example: "The HPV vaccine is comparable to the vaccine Pandemrix against swine flu"; "The HPV virus causes more diseases than just cervical cancer"; and "Of course you can get vaccinated against cancer". It also includes assumptions about sex, such as: "Most people find it embarrassing to ask about condoms" and "It's enough to use a condom to be protected against HPV" (Figure 2, page 83).

After you have responded to the five randomly selected assumptions, the app generates information about how many correct answers you had, and asks whether you want to play again. By classifying your performance as "bad", "fairly good", "good" or "very good", the app indicates whether you need to learn more, and, thus, need to play again. In doing so, the user's learning process is facilitated and mediated via the quiz. This articulates the quiz as an example of how the app is a care enabler. The app promises to facilitate girls' vaccination decisions through learning as play.

Several of the assumptions in the quiz are about sexual matters. As an answer to the statement "HPV is transmitted through air", the quiz generates the following answer: "No, HPV is transmitted through skin to skin contact and through sex, and the risk of being infected increases the more sexual partners you have". This articulates HPV as something that has to do with sexual bodily contact with other people, as generated between bodies through sexual relations.

The statement that the risk increases based on how many sexual partners you have, re-presents risk as something that can be more or less based on your behavior. Sex is staged as a "risk factor" that has to do with "risky behaviors". This includes a focus on sexual *lifestyle*; the message is that your risk level has to do with the way *you* live your life sexually.

In another example from the quiz, sex is present without a mention of HPV or cervical cancer. In this example it is stated that "Most people find it embarrassing to ask [their sexual partner] about using a condom". And the app generates the following answer: "No, it's the opposite! Studies show that 9 out of 10 think you are caring if you dare to ask about condoms". Thus, condom use is re-presented as a matter of caring for your partner. Making sex present, as is done in this example, without mentioning HPV and cervical cancer, is otherwise not common in the HPV vaccination context.

Condom use is however brought up in one of the other statements in the quiz: "It is enough to use a condom to protect you against HPV viruses." The answer goes: "No, a condom does not protect you 100 percent since the virus can be on body parts other than where the condom is placed. But condoms give some protection and protect against other diseases, too". Condoms are, thus, articulated as something that may protect you, but not to 100 percent. In contrast, it is implicitly asserted that a combination of vaccination and protection is needed. Condom use combined with HPV vaccination is presented as the way to go. How sex is made present in these examples differs from how HPV vaccination campaigns normally seem to sideline sex in favor of cervical cancer (see e.g. Mamo et al. 2010; Charles 2013, 2014).

With the focus on condom use, sex is not made present as a matter of anticipating "future sexual activity" (Lindén 2013b: 91). It is rather assumed as an important part of teenage girls' current life and *therefore* both HPV vaccination *and* condom use are needed. Even if this normalizes sex as a regular part of teenage girls' lives, it articulates sex as first and foremost a matter of risk. Instead of being a matter of potential pleasure, desire or happiness, sexual activities are made present as a matter of "sex negativity" (that is, as risks).

The quiz is related to a "gamification" of health data, through which "'playful frames' are applied to 'non-play spaces'" (Rich and Miah 2014: 310). The app turns information into a game one can do again and again until one *gets the facts right (*at which point the app qualifies the performance as "very good"). The app's game is a learning device in how it values users' performance. Following this, *the HPV app is an app, but it is also a game.*

An important difference between the HPV app quiz and other cases of apps discussed as gamification (see e.g. Rich and Miah 2014; Till 2014; Lupton and Thomas 2015; Maturo and Setiffi forthcoming), is that the quiz in the HPV app does not accumulate your data and personalize your use of the app based on this. The quiz can be accessed offline and the quiz data *is already there*. It is a learning device, with data already set. The quiz does not share results with third parties, and users' results are not connected to flows of big data.

"How good is the HPV app?": a questionnaire

In the last section of the app ("help us!") the user is asked to answer a few questions. Together with the three push notifications, this feedback section is one of the two parts of the app where *new* user data is generated. However, this data is hardly "big". This data is sent to the county council (to my interviewee Stefan's email); that's it.

The information gathered from the feedback part of the app is the user's age and region as well as an answer to the question "how good is the HPV app?" (choices range from "bad" to "really good"). The user is also encouraged to write a comment at the end (but this is not required for submitting feedback). The majority of the feedback messages (there were 40 in total when I accessed them in the spring of 2013) merely contain info about age, region and the user's rating of the app. Thus, they include no comment from the user to the county council. The feedback messages that do include comments, however, are diverse.

Some of the comments articulate worries and concerns. For instance, one user writes that "all the facts are not correct, it only protects against some HPV [types]" and another writes that it is "hard to understand the point with the format. With so much text-based information it would have been better with a web page which would perhaps have generated bigger diffusion."

Other comments are more positive. "Thanks for a good app. It helped me dare to take the shot", one user writes. Two other users write that "This has helped me a lot" and that "There are things I didn't know and I learnt more about things I did know". Yet another writes: "Really good that young people get this information. An app is the best way to reach out to us teenagers".

These comments give some insights into how users relate to the promise of the app as a care enabler. The user that brings up that the app is perhaps too close to a web page, invokes hesitations concerning the usability of the app. Since she is worried that the app is too close to a regular web page, a concern about how well it works as a care enabler is articulated. In contrast to this, the more positive comments are different. Emphasizing that the app has been helpful, that it is good that it provides needed information and that an app is the best way to reach girls, these comments can be read as alignments with the promise of the app as a care enabler. In turn, the critical comments can be understood as acts of *distancing* from the promise of the HPV app as a care enabler. These critical comments exemplify that the users of the HPV app did not simply conform to a vision of the HPV app as a care enabler, but that they also questioned whether it holds its promise.

The feedback function in the app can be related to a long history of enabling and allowing patients to provide feedback through feedback questionnaires. Whereas patient feedback is often sent via the post, the app allows feedback to be sent directly via email to the county council. It is a digital, app-located, questionnaire. The feedback function in the app is both app-specific, and not. *The app is an app, but it is also a questionnaire enabling feedback from users.*

Conclusions: caring for a very simple app?

It is time to go back to where I started this chapter: with care. I have cared for the HPV app in terms of a "care for neglected things" (Puig de la Bellacasa 2011). In doing so, I have explored how attention to the HPV app, as an example of what Lupton (2014: 608) defines as "very simple" apps, can facilitate an alter-narrative to big data and self-tracking ones. More concretely, then, what narrative has been enabled and generated by my care for the HPV app?

First, the HPV app is not that different from many other health communication devices. Through for example, its "did you know that ...?" and "why get vaccinated ...?" parts, it is a device that *provides information* in a manner similar to many other health campaigns. Following this, I have argued that the HPV app is an app, but it is also a movie, a pamphlet, a regular web page, an encyclopedia, a "to-do" list, a paper document, a questionnaire, a push notification device, a calendar and a quiz. These are "refashioned" in the app, and they *assemble* and *enable* the app. Many of these devices are, what often is articulated as, "older" ones. This makes the HPV app a device that does not easily fit an assumption that digital devices are radically different from "older" devices. Rather, to a high degree it inhabits, and is enabled and assembled by, *temporal consistencies* between "the new" and "the old".

There are parts of the app that make it different from "older" health communication devices. I focused on how the "push notifications" can be understood as a specific form of push responsibilization. In sending reminders about taking the shots, girls are encouraged to care for themselves *through* vaccination. Sending the information directly to the girls to their mobiles, something like a push responsibilization is performed, which first delegates care responsibility to the app (it takes care of remembering when to take the shot "for you"), and then back to the app user. Likewise, enabling users to generate feedback directly via an app, instead of by, for example, answering a questionnaire sent by regular post, is different. Additionally, as I have argued, the quiz can be understood as part of an increased "gamification" of health information.

I have also shown how coexisting media devices make up the HPV app. This sheds empirical light on how "media is always other media" (Kember and Zylinska 2012: 19); how media always is mixed media and that "the 'content' of any medium is always another medium" (Bolter and Grusin 1996: 339). The app is an app, but its status as an app is enabled by its coexistence with *other* media devices, devices that assemble the app into what it is. As an app that inhabits coexistent media devices, it is enabled by multilayered temporalities of mediation. It is made up by "newer" health communication devices, but, simultaneously, also by "older" ones. It is a "newer" device yet *it is not*. Based on this, it is possible to problematize the very idea that the app is a case of a "very simple" information app. *How* is an app that is made up by such temporal mess very simple? Is it so just because it is not big data? Can such an assumption hold?

Based on the analysis of the app, I argue that it is a problem to assume that the HPV app would be very simple only because it is not complex *in the same way* as big data apps. It is rather complex in a manner that does not fit a technoscientific "hype" about big data and self-tracking. Being a device that is enabled by a range of coexisting devices, it is "more than one, but fewer than many" (Mol 2002: 84); both singular and multiple.

Following this, I conclude that *my care* for the HPV app as a case of a "neglected thing" has enabled a different story about smartphone apps. This story troubles the assumption of not data-generating apps as very simple by telling an alternative story about *another version* of digital mess and complexity. It tells a story that has to do with coexisting temporalities of mediation that entangle "the new" and "the old". My care troubles a technoscientific hype, and what it enabled can serve as an important alter-narrative that enables other engagement with smartphone apps. Thus, my care for the app as neglected has enabled and generated a troubling of temporal simplifications, and has pushed for a need to further stay with the trouble of apps.

However, my care for the app as enabling an alter-narrative is not done in a vacuum. There is more trouble here. As Puig de la Bellacasa (2015) stresses, alternative care temporalities are not outside the politics of the predominant ones. This evokes a need for attending to additional dimensions and implications of *my care* for the app. As I have shown, the HPV app does not stand outside politics of care already circulating. As a device that, even without my care, is a care device and a care enabler, my care for the HPV app *is* entangled with the versions of care the HPV app *already* inhabits and enables. When I care for the HPV app as neglected, I also "automatically" care for the versions of care already circulating in, and through, the app. I cannot turn the HPV app into a matter of care; it *already is* a matter of care. What I have done is added a further layer of care.

This makes it important to discuss the dimensions of care that I "automatically" care for when I care for the app as neglected. If the app is a care enabler, my care for the app as neglected simultaneously feeds into a care for it as a care enabler. But what care does it enable? And *for whom*? For whom, on the contrary, could it be troubling or even harmful? Puig de la Bellacasa raises related concerns. Learning from her, caring for marginal things "is never a neutral affair" (Puig de la Bellacasa 2014: 38) and it is important to ask "for whom?" (ibid.: 29) and, thus, "for whose benefit?" (ibid.: 38). Also Eva Giraud and Greg Hollin (forthcoming: 6) share similar concerns. By quoting Haraway (2008: 87), they stress that theoretical engagements with care need to ask "for whom, for what and by whom". As the HPV app is a device for public health governance, these questions are important.

As I have shown, the information mediated via the app includes statements about sex, protection, cervical cancer, and HPV vaccination shots, which are closely linked to assumptions about gender and sexuality, and a vision about neutral, objective facts "from nowhere". For example, sexual activity, I have discussed, is made present in the app *as a risk* for sexual infections, and the presence of the caring school nurse and the feminine-looking girl subjects comes with gendered tropes about the app user, and about the nurse performing gendered care labor. Therefore, the versions of care the app inhabits come with specific knowledge, gender and sexual politics that are closely linked to stratified promises of public health, including ideas about *who* the HPV app enables care *for*.

Following this, the app does not just enable *any care*. It encourages a girl-centered self-care *through* commonly circulating gendered re-presentations of girls and women. In this way, the app is part of a gendered politics of care. That is, the app enables care *through* tropes linked to normatively oriented gender and sexual politics. In this way, how care is enabled by the app is a clear example of the politics of care. No matter how important it might be, a care for the HPV app as something that enables an alter-narrative to technoscientific "hypes" cannot be imagined as standing outside of existing politics of care, and which might include for example a gendering of care. Accordingly, when caring for neglected things, "for whom, for what and by whom" certainly need to be asked.

Following this the app. does not provide any [illegible] outcomes, [illegible] [illegible] [illegible] [illegible] [illegible] [illegible] [illegible] [illegible] [illegible] [illegible] [illegible] [illegible] [illegible] [illegible] [illegible] [illegible] [illegible] [illegible]. That is [illegible] [illegible] [illegible] [illegible] [illegible] [illegible] [illegible] [illegible] [illegible] [illegible] [illegible] [illegible] [illegible] [illegible] [illegible] [illegible] [illegible] [illegible] [illegible] how important it might be, [illegible] the [illegible] app is something [illegible] [illegible] [illegible] [illegible] [illegible] cannot be [illegible].

5. Facts, Fears and Friction

"There's a lot around vaccinations, many that warn you ..." my interviewee, the county council communicator Hanna, told me when I met her at her office to talk about her work with the HPV app. Due to this, she continued, "we wanted to communicate information that we knew there was a lot of difference of opinion about". Hanna, and the other health professionals in the HPV app working group at Mittland County Council, stressed in the interviews that the information in the HPV app hopefully helps girls to make a decision about whether they want to get vaccinated or not. They emphasized the app as communicating neutral, impartial and scientific information, and contrasted this with rumors, stories and fears. In doing so, these interview narratives partly resonate with a vaccination landscape in which "science-as-epidemiology" is often contrasted with worried citizens whose actions are understood as driven by feelings and based upon unreliable experiential knowledge. Accordingly, Leach and Fairhead (2007) state that in the context of vaccinations "rumor has become a shorthand for an idea that can be replaced with proper 'facts'" (ibid.: 33). This distinction between rumors and facts builds upon the "assumption that rumors will abate with 'proper' biomedically oriented scientific information" (ibid.). Indeed, how my interviewees articulated science-as-epidemiology as a "clear-cut solution" (Jerak-Zuiderent 2015: 425) to vaccination fears and myths, in some ways fits aptly with other discussions of vaccine anxieties and fears in contemporary society.

I study how the HPV app working group discussed vaccination information in interviews, including its relation to the HPV app and to girls. I deepen the discussion from Chapter 4 on the content and materiality of the app by paying attention to how the app, and the information in it, was reflected upon and discussed by the working group.

Feminist STS scholars, such as Jerak-Zuiderent (2014) and Puig de la Bellacasa (2015), argue that paying close attention to *moments of friction* can foster (more livable) matters of care. Inspired by this, I draw upon Jerak-Zuiderent's careful attention to moments of friction *within* otherwise coherent interview accounts. This entails a need for slowing down the analysis to allow space-time for subtle moments of friction that are easily brushed over. More concretely, based on inspiration from her study on how laughter can potentially be a generative matter of friction, I will attune to (marginal and momentary) disruptive laughter, hesitations and silences as a way of slowing down HPV vaccination communication matters. As Jerak-Zuiderent (2014: 904) shows, this includes the importance of listening closely for "fleeting subtleties" as a possibility to fostering (more livable) matters of care.

Resonant with Jerak-Zuiderent's study, my interviewees reflexively discussed their work by invoking science-as-epidemiology as a clear-cut solution to vaccination fears. However, by slightly shifting focus and slowing down the analysis, I will show how many less clear-cut articulations were circulating *within* the (seemingly coherent) expert narratives. These less clear-cut moments, I will show, disrupt a meta-narrative told about science-as-epidemiology. As I will show, these moments enable a focus on certainty as something fleeting and contingent, and uncertainty as always, already present within the coherent or certain. Accordingly, my aim with this chapter is to point toward moments of friction as a way to loosen some of the tighter knots presenting science-as-epidemiology as *the* solution to vaccination fears.

I begin with a discussion of how the HPV app was invoked as a good and fitting device for the job as it was believed to enable the county council to reach girls where they are. This will be followed by attention to how this narrative was troubled by disruptions, hesitations and uncertainties circulating *within* the interviews. From there, I will move on to how science-as-epidemiology was articulated as certain and stable "good information", and was contrasted with vaccine fears, myths and rumors. The chapter will facilitate an unsettling of (seemingly) clear-cut visions of "good information" by attuning to fleeting subtleties that provide alternative visions. Finally, I move on to how a *view from someone* was envisioned as generating trust, something that complicates science-as-epidemiology as a *view from nowhere*. I end with a discussion about what this attention to moments of friction, fleeting subtleties and alternative narratives illuminate regarding a discussion on matters of care in technoscience. In particular, I will relate this to a discussion of coexisting temporalities of care.

Reaching the girls where they are as a matter of care

The information secretary Katarina said in the interview that they needed to find a way to "communicate where [the girls] are in life" and that girls often communicate "via social media and technology". Katarina, and several of my other interviewees, saw schools as the best place to reach girls. But since the catch-up vaccination is part of the care choice system, girls are not automatically given information in school (as they are in the context of the regular vaccination scheme). Also, since the schools in Mittland, as mentioned in Chapter 1, did not decide to authorize as vaccinators, they were outside the catch-up vaccination scheme. Therefore, my interviewees emphasized that it was important to reach girls at other places where they were likely to be found. The gynecologist Karin said:

> I think the idea and thought behind the app is to try reaching teenagers in for example, their territory or their arenas. As well as information on Facebook and so on, we thought an HPV app would be good as you yourself can click to get the information and get answers to the questions you want.

Karin said that the app made it possible to reach the girls in their arena or territory, and explained it as a device that "fits our time" due to *how* it communicated HPV vaccination information. This is reminiscent of how others of my interviewees talked about the issue. Sara the school nurse stressed that "it's very much a part of our time that you download apps" and the communicator Hanna said that "we tried to use what we thought … where they are". In this vein, the app was stated as a suitable communication device for reaching girls as it reaches them *where they are.* Since the girls could not automatically be reached at school, my interviewees emphasized the app as a digital arena as a good solution.

This has to do with care. The app was envisioned by Katarina, Karin and several others of my interviewees as a device that would enable better care as it locates care to where the girls are. As described in the introductory chapter (Chapter 1), due to the fact that the catch-up vaccination is a part of the care choice system, girls need to find a vaccinator and decide to get vaccinated. In the interviews, providing good care got translated into making it easier for girls to get the information they needed to make a decision about whether they want to get vaccinated, or not. Therefore, *reaching girls where they are* was articulated as an important matter of care, and the app as a device that enabled this care.

The information secretary Katarina also related the wish to be where girls are to bigger questions of increasing or enabling public participation. According to her, it was important to work with the public concerned or, at least, finding ways of relating to what that public needs. Therefore, it was also crucial to enable communication technologies that reflect the public's need. In her words:

> I guess I'm thinking about your whole area around technology and the relationship to the social ... It's humans we're talking about and I think it's really important at all times to be close to where humans are in their lives and how they are living their lives and if we come close to that I think we will always achieve more and it's, sort of, to adjust the technology to the context so that it fits different societal structures and the specific conditions people live under and things like that. So that's, you know, an important part we should all think about becoming better at in all possible situations; to increase participation.

In Katarina's reasoning it is important to find technology that fits a specific way of organizing society, including the societal conditions that affect one's life. The app was envisioned as a technology that fitted a specific part of contemporary society: teenagers' world. Care for girls was envisioned as improved by the app as it gave girls a space or arena *through* which they could participate. Katarina's argument was even more broadened to include human life in general and not only the specific example of HPV vaccination. She stressed that it would be possible to get closer to lives of humans if public participation was improved. During the interview, Katarina repeatedly emphasized that people – girls included – were active and wanted to participate in matters that relate to their life. This idea of girls as active citizens or participants is a common trope in a context of HPV vaccination (see Lindén 2013b).

The idea that including people in public matters would generate technology that fits the whole targeted population builds upon an idea that a few people from the group can re-present the interests and desires of the whole population "out there" (and as such, it is a "representationalist" trope). By drawing upon an assumption that humans are active and want to participate in public matters, Katarina thus invoked an universalist trope assuming humans to be active citizens.

Evident in Katarina statement that it is important "to adjust the technology" so that "it fits", is that she includes an idea of technology – such as the app – as something that needs to be adjusted to serve human needs. For her the app is a technology (or a device) that is meant to serve a specific purpose, namely to serve the interests of girls. In the quote from Katarina, as in other interviews, the app is articulated as a fitting care enabler as it works as a *transmitter* of the HPV vaccination information necessary for girls "out there".

Katarina stated that the quiz in the app communicates HPV vaccination information in a playful way that makes it more fun. With the app, "girls can, you know, get the information through play" and that, she added, "makes it more fun". Here, the app (and the quiz as a part of the app) is not envisioned as changing the information provided/transmitted. Once again, it just transmits the information, albeit in a more fun and playful manner. Thus, the app should not transform what information is, or can be, communicated. The information is already there; it just needs to be transmitted in a way that suits the girls concerned. The county council merely needed a good tool for the job; a device that could effectively transmit the existing information to girls, and which could provide this information to girls where they are. This idea that the information was already there makes absent how the app (as I discussed in Chapter 4) comes with specificities for *what* and *how* HPV vaccination is mediated.

Troubling the app as a good device for reaching the girls

However, it is not quite that simple. When I asked the school nurse, Sara, "why an app is suitable for communicating HPV vaccination information", she started by tapping into the narrative told by, for example, Katarina: "it's part of our time that you download apps". Then she hesitated, and was silent for a bit. When she continued, she compared the app to the bio advertisement movie for the app[1] and answered:

> At the same time, I know this, as I have teenagers myself, that they have very limited space on their mobiles and they don't download apps they don't want [...] The idea is interesting but I can't say that this will be the big thing for the future. But I do notice that what actually did attract attention is really this cinema advertisement. But perhaps I'm not allowed to say that here [laughter]. But I do notice that. It has generated a huge response and attention, that short film. And when I've heard discussions where this has been mentioned ... Sure, the app is a part of that but this cinema advertisement from the county council ... Short, stylish. It felt, you know, as a very interesting form to actually use for other work as well. But perhaps that isn't what you want to hear [laugher].

1. When the HPV app was launched, it was marketed via a cinema advertisement movie where the school nurse re-presented in the app encourages the public to "download my app". It was screened before the Twilight film in 2012, as the county council anticipated that this film would attract the girls being targeted.

By first hesitating and being silent for a moment, and then laughing and saying that I perhaps do not want to hear her saying that the app is not that suitable, this part of Sara's and my conversation says something about what Sara anticipated me wanting to hear – and perhaps what she as a part of the HPV app working group felt that she should answer. Being asked about why the app is suitable, my question led to hesitation, silence and laughter, and later seemed to evoke a need to make excuses for feeling that she could not answer the question in "the right way". I will return to Sara's laughter, hesitation and the silence later, after I have included some empirical material that can further deepen an understanding of the importance of these quite subtle moments in the interviews.

Sara's way of answering the question says something about the app: the app was not simply articulated by her as a good device for the job. Instead, apps, in Sara's words, are perhaps not "the big thing for the future". Through the use of experiential knowledge Sara, as seen in the quote above, made use of her position as a parent and school nurse to problematize the app. In doing so, she argued that the thing that took off was not the app but the advertisement about the app. Therefore, she concluded, perhaps cinema advertising is, in fact, a better medium for reaching girls *where they are*. If girls do not download the HPV app, it does not serve its purpose. Accordingly, in Sara's answer the app is articulated as a potentially unfitting device in how it was feared to be outmaneuvered by other more exciting devices. Invoked as an imperfect device for the job, Sara indicated that the app perhaps does not have enough attraction for teenagers, as it is not something they download, as she said, "for fun". In contrast to Katharina's emphasis on the app as enabling playful learning, it was feared by Sara to not make learning *playful enough*. Feared as unfitting, and invoked as imperfect, the app was circulating in this part of the interview as a potentially *failing* matter of care.

When being asked the same question as Sara (why the app was suitable for the task), the communicator Hanna hesitated in a similar vein, was silent for a moment and then said: "Yeah, why do I think the app is needed?" She then continued by talking in general – not HPV app specific – terms about why information is needed. After having talked generally about information for a while, she ended her answer by stating "but this app specifically ... why an app is needed, that I cannot really say". In this way, the HPV app was not discussed as *the* solution for reaching the girls. While Hanna stressed that information is important, the need for information mediated via an app was evoked as less clear, and more uncertain. This uncertainty impelled hesitation and silence. And when not being able to answer the question about the app in a coherent manner, Hanna moved on to talk about information in general.

The hesitation both Hanna and Sara expressed (and the silence that allowed space for it), as well as the burst of laughter from Sara, do not quite fit an interview practice of clear answers. The disruptions of hesitation and laughter hint at a moment of friction between an anticipated answer in accordance with a meta-narrative (of apps as part of our time, and of the HPV app as a good device for communicating needed information) and at something that does not quite fit. This articulates a friction between a vision of the app as the solution for *reaching girls where they are* (and thus, as a care enabler) and its functionality as quite unclear and uncertain.

In a context of feminist STS engagements with politics of care in technoscience, Schrader (2015) hints at the possibility of attending to hesitation as a fostering of matters of care. In the same way that I wrote in the introduction to this chapter that Jerak-Zuiderent (2014: 898) does, Schrader argues for a generative potentiality of slowing down things. This, she asserts, can enable a space-time for the unexpected, as well as – and this is of direct relevance for this chapter – for "the generation of space-times for *hesitations*" (Schrader 2015: 684, emphasis added). This argumentation, I read as in close affinity with Jerak-Zuiderent's (2014) attention to laughter as a "fleeting subtlety".

Relating this to Sara's and Hanna's hesitations, and Sara's burst of laughter, it is possible to see how these subtle moments have importance. As moments of disruption in a meta-narrative about teenagers' use of apps in contemporary society, Sara's and Hanna's answers make present an alternative vision of the app. Interpreting these as moments of friction shows that they did not invoke this alternative vision as clear-cut and outside the current dominant narrative. The hesitations and laughter were moments of uncertainty and unease, not clear-cut articulations of alternatives. Instead these moments *rubbed up against* the meta-narrative of the app as fitting, and disturbed it *from within*. Based on this I argue that staying with these marginal moments makes it possible to see how these fleeting subtleties trouble a given status of the app by doing so *from within* the otherwise often coherent narratives.

Later in the interview, and talking in more explicit terms about her hesitation, Hanna said:

So perhaps apps, at least this form of apps, lose a bit of their function, as you can do these things on regular web pages. But that of course, apps as games and apps that have another kind of functionality with small programs and things that really do something ... But this one doesn't. This is more like information and a quiz and things like that.

Not being an information app that generates user data, the HPV app's usability was questioned. It was feared to be *too close* to regular web pages. The HPV app was expressed as a device that does not fit a narrative about how new and different apps are.

In a similar way to how Sara invoked the app as imperfect, in the interview with the information secretary Katarina, and the doctor Stefan, the app was discussed as a device that existed in a situation of "many possible roads". Moreover, Katarina stressed that "we tried to look into apps and how they work and then somewhere you understood that they might disappear soon". Stefan added to the conversation by saying that "[i]n a year it's not an app but something else, a different combination of letters", indicating that apps may be outmaneuvered by other communication technologies. Similarly to how Sara talked about it, the app was not stated by Katarina and Stefan as a given decision or solution. Even if they did not discuss the app as imperfect in the same way as Sara (and did not show the same hesitation), they did not state the app as *the* device for reaching the girls.

In different ways, Hanna's, Sara's, Stefan's and Katarina's answers (including the laughter, hesitations and silences) problematize a narrative of the HPV app as a given solution for mediating HPV information (and care) to girls *where they are*. If girls do not use the app due to it being not "fun enough" or too similar to "regular web pages" it fails as a care enabler. This shows that *reaching girls where they are* as a vital matter of care was not envisioned as *taken care* of once-and-for-all through the app. Instead it remained an open question whether the app actually worked as a care enabler.

Absolute certainties and bursts of laughter

[I]t's very important that we stand for correct information [..] [L]ast spring there were so many rumors, myths, stories. You could even read that if you take this vaccine, then you will lose your arm! [...] In essence, there's no component at all that's the same [as the swine flu vaccination]. This is tested on a lot of girls. In essence, we have no big side effects [...] People found a few cases of deaths after having taken the vaccination. And then it's about looking at this medically. Sure, there's actually like three cases registered [...] But there was no link indicating

> that they had died due to the vaccination. That we had to look at. And, unfortunately, people die sometimes [laughter] [...] One of our most important roles is to be impartial in this. We should not hide the fact that sure, it may hurt, you may get a swollen arm. That is, things that *actually* can happen to you.

In this quote, health care planner Johan stated that information communicated by the county council needs to be correct and impartial, something that is translated into a need for providing scientifically "approved" information to enable girls to make a vaccination decision. By stating that it is about looking at possibilities for side effects medically, he implied that this was the way to handle the matter.

Johan stressed rumors and myths (such as that one might need to amputate one's arm, or that people are dying). In doing so, other possible questions, worries or fears girls may have concerning the vaccination were made absent. Additionally, that he stressed a swollen arm as an example of what *actually* can happen to girls, a possibility for uncertainty around HPV vaccines was made absent. People can get a swollen arm from the vaccines, they do not die. That is a given. In the same way, stating that "and, you know, people die sometimes" (and thus implicating: but not from HPV vaccines!), HPV vaccines are articulated as safe. The main lesson from this reasoning is that side effects are nothing to worry about since HPV vaccines are well-tested, and therefore safe. In this way, Johan's answer echoes a trope assuming "proper" or "good" vaccination information to be science-as-epidemiology. How to treat stories, myths and rumors about side effects becomes a quite simple problem with a clear-cut solution: just respond with medical facts. Here, science-as-epidemiology becomes a matter of care in how it is envisioned to counteract rumors, myths and fears, and, therefore, enable HPV vaccination care.

Nevertheless, two quite subtle things happened in the quote above. First, Johan added twice that *in essence* there are no big side effects. That Johan seemed to feel the need to include *in essence* opens up uncertainty and that HPV vaccination may not be a given good. Importantly, this illuminates a lingering (always, already) presence of uncertainty that seems to prompt Johan to include "in essence" as a precautionary wording. The second thing that happened was that Johan laughed briefly after having said that people sometimes die. This short disruption of laughter can be interpreted as not only indicating death as a generally uneasy topic, but also as suggesting that Johan finds the idea impossible that HPV vaccines would include a component of uncertain safety. His laughter indicates that people do (unfortunately) die – but that it is not possible that they would die of HPV vaccines. He laughed at what was articulated as *alter to* the certainty of vaccine safety. However, the fact that Johan seemed urged to emphasize the safety of HPV vaccines and "laugh away" possible death and uncertainty when responding to stories about side effects, also points to an always, already present uncertainty. By "laughing away" uncertainty, certainty is confirmed and secured. By drawing upon Jerak-Zuiderent's (2014) analysis of the relation between laughter and fear, it is possible to speculate that Johan here seemed to laugh at, and fear, the same thing: (death due to) uncertainty.

Also Hanna stressed, when talking about how the swine flu vaccination has generated public fear for other vaccines (as the swine flu turned out to cause narcolepsy), that "this fear one can understand and is a fear one needs to take seriously and, so to speak, still respond to with facts". As Johan did, Hanna contrasted facts with fears. In both interviews, facts were envisioned as the solution to fears. A common binary dichotomy was articulated: rational science versus stories, myths and feelings (such as fear). During these moments of the interviews, this view from nowhere version of knowledge closed down the possibility of taking seriously matters of care that did not fit the distinction between good information and mere stories. Instead, scientific information was articulated as "winning over" stories, myths and fears (and science was separated from storytelling). In this way, facts were articulated as a clear-cut solution to the problem of fear.

The contrast between fact and fear was evident also in the interview with Katarina. She explained the problem with fear as having to do with a lack of knowledge. "Without knowledge you don't have the capacity to make a difference between vaccination and vaccination" and "it's often a lack of knowledge that is behind these concerns", she said. Reminiscent of how Johan brought up rumors of people dying due to the vaccination as examples of storytelling, Katarina brought up reading on a website and reading that a girl has died as an example of fear-based – and not correct and substantial – information. She stated that it was important that "[you] get the right knowledge to base your decision on so that you don't go onto a website and see that a girl died in the US … that you get scared and make up your mind based on that". She thus assumed that fear is a problematic basis for choosing whether to get vaccinated or not. And due to this being a problem, good science-as-epidemiology information is needed.

For Katarina an important part of the county council's communication work was, as she said, "to increase the possibility for [girls] to get correct information and substantial knowledge to enable a decision". Envisioning fear as based on a lack of knowledge, "good information" (facts) is believed to enable a rational decision. This decision, she continued, "doesn't have to mean that they decide to get vaccinated", as she has "complete respect for a decision to refuse it". However, even if girls decide to refuse the HPV vaccination, she still imagined that "good information" would lead to a rational decision.

This assumption of a lack of knowledge as the reason for people being scared or hesitant about vaccinations is often invoked in vaccination policy. As Leach and Fairhead (2007) stress, this assumption tends to make absent "forms of experiential expertise grounded in everyday practice, knowledge and epistemology" (ibid.: 23). Hence, it tends to make absent reasons for why people are scared of, or hesitant toward, vaccinations other than lack of knowledge (additionally, it builds upon assumptions about what counts as knowledge and what does not). If fear is about a lack of knowledge that can be counteracted by factual science-as-epidemiology, the solution to the problem is simple: educate the concerned public (educating here meaning providing "impartial" and "correct" information).

This idea that facts will solve the problem of fear includes the idea of a subject that, when being provided with factual scientific information, will stop acting based on fear and start acting rationally. More information will abate fears and generate a rational decision. Thus, responding to fears and worries by counteracting it with facts envisions a scenario where more *good* information will (as a linear, progressive movement) lead to people making a *rational* decision. The invoked dichotomy between science and feelings, therefore, articulates a situation where affective responses to vaccination are taken seriously *through* an assumption that girls, if just given the facts, will stop reacting affectively, and will start acting rationally.

A contrast between fears and facts was invoked by Hanna, as well:

> [HPV vaccination] is a good thing. And then it's a shame if you're scared of a vaccine that protects against a severe disease, now when there's protection. Then you want to do everything to change their minds and make them dare to get vaccinated. You should not refuse because you're scared or worried or that for by different reasons you feel that you are scared of this [...] But for me, it's important as a communicator at the county council that people make that decision

> based on impartial information. You should not need to make that decision based on the wrong premise or incorrect information. Believe in myths and sort of…

In this quote, Hanna says that it is a shame that people are too scared to get vaccinated, as HPV vaccination "is a good thing". HPV vaccination information (such as via the app) is needed to "change their minds and make them dare to get vaccinated". Johan, too, stated the vaccination as good: "as we have an efficient way of protecting 70 percent of cases of women developing cervical cancer, it is an absolute truth that we shall vaccinate as many as possible to get rid of this suffering". Here, an *absolute truth* effectively serves to position the safety certainty and efficiency of the vaccination as a given good, as a fact. In these examples, girls' fears of HPV vaccination becomes something that good information – here formulated as impartial information – can counteract.

For Hanna and Johan (this goes for my other interviewees as well), fear due to wrong information was emphasized as a problem as it decreased the possibility for girls to receive HPV vaccination (as Hanna stated, this "good thing"). Thus, for them, it is a shame that girls are afraid as they risk missing out on something that is good for them, something that is about care. It is worrying for them that problematic and biased information makes it trickier to reach out to girls and tell them about HPV vaccination as a "good thing". They need to do it *through* an emphasis on rumors as bad or false information. Invoking HPV vaccination as a matter of care, problematic information or myths becomes a matter of *concern*.

In contrast to Johan saying that it is an *absolute truth* that as many as possible should get vaccinated as the vaccines *do* protect against 70 percent of all cervical cancer cases, Hanna stressed the existence of side effects:

> [Y]ou should know that there are side effects but you also need to know that there's no reasonable health care authority in the whole world that puts that many resources … who vaccinates that many girls … if you don't know whether it's a safe vaccine or not […] [I]t's an individual decision if you choose to get vaccinated or not. You perhaps feel that it's going to be okay anyway. And it may. The majority don't get sick with cervical cancer, which is the case with all diseases. That goes for other vaccinations, too. Side effects exist.

In this quote, for Hanna, the existence of side effects makes it crucial to provide girls with good, impartial information that enables them to decide for themselves. In her words, the existence of side effects turns the HPV vaccination into an individual decision. However, at the same time, she emphasized that HPV vaccines are safe since health care authorities would not vaccinate so many girls if they were not. Nonetheless, saying that side effects exist implicates uncertainty about vaccine safety. These two examples from Johan and Hanna illuminate a tension between factual claims about safety and efficiency (the "absolute truth") of HPV vaccines and a lingering (always, already present) uncertainty.

In this section of the chapter, science-as-epidemiology as "good information" has circulated as a dominant matter of care. However, in conceptualizing Johan's laughter, his wording *in essence* and the articulation of a possibility of side effects (and of uncertainty) as moments of friction, I have highlighted moments that made present alternative narratives and possibilities that open up for uncertainty *within* an otherwise coherent narrative.

Uncertain certainties

When I and the school nurse Sara talked about a Swedish TV documentary program's (*Cold Facts*) episode about HPV vaccination, we discussed how they used a specific (very critical) angle. Talking about this, Sara invoked facts as not 100 percent certain:

> And I do feel safe [about the safety of HPV vaccines] as they have given so many doses and you can reduce certain forms of cancer [...] But I can reconsider that if other facts are developing. And you can always use a specific angle and skew [the facts]. And, once again, nothing is ... nothing is 100 percent.

Here, what good information is, is not a given. As facts are not 100 percent certain, they might be replaced by new facts. Good information becomes something that has to do with communicating *current* facts (statements that are seen as facts today). Similarly to Sara, Hanna stated that the county council has a responsibility to tell girls the facts they know *right now*. She compared HPV vaccination to an example of dietary methods. From our conversation:

> HANNA: But it's still our responsibility to tell the facts we know. But then we both know that facts change. Take, for example, this with dietary. You shouldn't eat fat and, you know, [you should eat according to] *tallriksmodellen*[2]. And for a

2. *Tallriksmodellen* ("the plate model") is a Swedish dietary model that builds upon the idea of it being healthy to eat 40 percent carbs, 40 percent greens and fruits and 20 percent protein (100 percent stands for the size of a dinner plate).

while we've been going for that. And now you notice that, what is this with you can eat fat but no carbs? That you can't … All of that you can't include.

LISA: Yeah, it changes.

HANNA: It changes. New facts are added. Then you have to take a new standpoint and that's life [laughter]. That's, you know, how it looks. We act based upon what we know today.

In our conversation, we agreed that facts are changing. In Hanna's words, this nature of facts, however, is not something that has to be included in public health information. She explicitly stressed that everything cannot be included. This serves to justify the presentation of HPV vaccination information as stable, given facts. At the same time, the changing nature of facts is in Hanna's words something that means that the county council needs to act upon "what we know right now". Similar to the quote above from Sara, facts are envisioned as somewhat *for the moment* stable and certain. This differs from assumptions such as it being an *absolute truth* that it is good to vaccinate as many as possible. Here, it is rather a temporally situated fact that is stable and certain during a certain period of time.

In a similar manner as with Johan's and Sara's laughter, Hanna laughed when stating that even if facts change you cannot include all these changes in health education information. For Hanna, the fact that facts are *for the moment* stable and certain means that you as a representative from a governmental institution need to take standpoints based upon what you know at the moment. By laughing when saying this, however, Hanna's answer indicates a fear of not taking the right standpoint and, therefore, giving people the wrong information. This moment of disruptive laughter can be speculated as being a moment that indicates what Hanna might fear. Hanna acknowledged uncertainty ("facts change"), and seemed to fear it at the same time. The facts she talked about inhabit what can be discussed as an *uncertain certainty*, and can be speculated as being what evokes a fear of making the wrong decision.

The articulation of *for the moment certain facts* (uncertain certainty), says something about matters of care. In communicating what is for the moment certain, a matter of care other than science-as-epidemiology as a clear-cut solution is staged. According to this articulation, *reaching the girls* where they are with good information as a matter of care requires careful attention toward a possibility of uncertainty and/or of change. Instead of staging that things have "an absolute truth" as Johan did, ongoing attentiveness is required toward the fact that facts change and differ. As Hanna stated, "that's life". This means that it requires the county council to *live with* an attentiveness toward the always, already presence of (potential) uncertainty. It requires them to work with contingent and changing accounts of good care and an always, already present "living with uncertainty" (Jerak-Zuiderent 2012: 736).

However, by not making the uncertainty and contingency of facts present in their actual campaign devices (such as in the app), the county council retains such openness and contingency toward what counts as good care within its own boundaries. The idea that girls need to be reached *where they are* as they want to participate in, and contribute to, matters *they* care about, becomes in the formulation of Hanna a highly conditioned practice. As the "good information" is already there (even if it may change over time), girls can only participate and be active insofar as they do not disturb the *public presentation* of facts as "an absolute truth" or *certain certainties* (and not as *uncertain certainties*). Therefore, it becomes a highly conditioned matter of care that is tightly connected to public participation as a mode of governance that makes absent the uncertainties of the practice.

Generating trust through facts (from someone)

In the interviews, vaccination fears were often linked to how narcolepsy turned out to be a side effect of the swine flu vaccination. Why did my interviewees imagine that the swine flu vaccination affects HPV vaccination information practice? The health care planner Johan answered: "In school you've always been given your vaccines. But here you need to go there yourself in the same way as with the swine flu vaccination. Therefore, I think there's a fear for that". In this vein, the decision to go for a catch-up HPV vaccination outside of the school health system was explained by Johan as a reason for why people related the swine flu vaccination to the catch-up HPV vaccination. Other interviewees emphasized that girls were scared of HPV vaccination, as teenagers were the ones who developed narcolepsy after having taken the swine flu vaccination (and the ones especially targeted by the swine flu vaccination as it was assumed that older people would already have developed resistance). Consequently, Hanna said: "We have, you know, had this vaccination against the swine flu where it was exactly teenagers that were harmed and got narcolepsy. And that has generated a fear for this vaccine, too, as it involves the same target group as the group that got sick with narcolepsy". In a related vein,

Stefan emphasized that even if there is no connection between HPV vaccines and the swine flu vaccine, "people make that connection, still today people make that connection".

This emphasis on how the swine flu vaccination affects the catch-up HPV vaccination was often related by my interviewees to what was mentioned as a problem of trust. For example the information secretary Katarina said:

> It feels like the county council, or the National Board of Health, has suffered a severe blow to trust due to the swine flu and that vaccination […] Since here we have – even though it could be a mini mini mini chance that something will happen, it always is – a highly well-tried out vaccine, compared to the other [the swine flu vaccine], which was developed really fast. So naturally, it's often lack of knowledge that is behind these concerns […] It's probable therefore that many don't get vaccinated. You don't dare to trust that this is best for you.

As already discussed, Katarina connected vaccination fears to a lack of knowledge. In addition, as brought up in this quote, she connected vaccination fears to a problem of girls not trusting the safety of the vaccination. This was stated by her as one of the reasons for why girls did not get vaccinated (and why they were scared), and was being related to a lack of trust caused by the swine flu vaccination. Hence, a problem with lack of trust was associated with a lack of knowledge (about the safety of the vaccination). People lack trust if they lack knowledge. This envisions an idea that more knowledge (more "good information") will make people trust the vaccination. In other words, and as partly already discussed in this chapter, facts were envisioned to abate fears, and to generate trust.

In the quote above, Katarina stated that there always is a "mini mini mini chance" that something will happen. This opens up a possibility of vaccination uncertainty. Thus, Katarina (in a similar vein to how I earlier talked about how Hanna and Sara stressed facts as changing and Johan stated that there is *in essence* no big side effects) did not say that there is an absolute truth that HPV vaccines cannot generate side effects. However, stating that this chance is not only little but *mini mini mini*, it is not raised as a reason for being worried. It is not a reason for not trusting the county council and the vaccination. In other words, it is not a matter of concern that needs to be raised or extensively discussed. On the contrary, she stressed that Gardasil was a highly well-tried vaccine. An important matter of concern for Katarina was that bad information makes people stop trusting the safety of the vaccination. Due to this lack of trust, good information that *generates* trust is needed (and needs to be communicated to girls where they are), and becomes a matter of care.

Katarina's emphasis on trust can be related to an often assumed "breakdown of trust" or "trust crisis" in vaccination practice (see e.g. Poltorak et al. 2005). However, such a problem of trust in the context of the county council's work with HPV vaccination was not stated by Katarina (or any other of my interviewees), or implied, as related to an idea of a *general* breakdown of citizens' trust in governmental institutions. A "blow in trust due to the swine flu" was, thus, not equated with a general situation of citizens not trusting the government and health authorities any longer. My interviewees did not assume, or argue, that the county council needed to rebuild trust relations with concerned publics in general. In this vein, my interviewees' statements differ from how a lack of trust in vaccinations is often discussed by researchers in vaccination research within social science, as well as in policies (see Hobson-West 2003, 2007; Brownlie and Howson 2005; Poltorak et al. 2005). This means that it is crucial to look at how trust comes to matter in *specific* ways, in specific vaccination situations (and also, how matters other than trust may be invoked as the important matters). It seems important to use trust in an emic and context-sensitive manner, rather than as a meta-framework for understanding vaccination resistance or rejection.

When I asked the school nurse Sara about why they decided to include a movie in the app featuring her as a school nurse she related this decision to the need for trust. From our conversation:

> SARA: You affect people in a completely different way if you use someone with expert knowledge and if you, kind of, get a face of someone who actually works with the questions. So, actually, I think it's a fairly good initiative [...] This generates trust since you know that the school nurse often has a close relation with the pupils and actually knows what it's about. So I think it was a wise choice to use a school nurse, regardless of whom, as it generates substance and trust.
>
> LISA: They trust more?
>
> SARA: Yes, I think so.

In this quote, Sara says that including a school nurse movie in the app was a good decision since getting a face of someone who actually works with the question tends to make people trust the information. Hence, it is not the information *in itself* that is judged to be trusted but information is stated as believed in when someone that girls trust communicates it. Hence, it is the connection between information and the school nurse that is stressed as generating trust. Accordingly, using a school nurse, Sara said further, was a good move as this person often has a close relation to the pupils. Thus, Sara envisioned trust as enabled when you know the person giving you the information.

The idea that the school nurse is a person girls trust was brought up in several interviews. For example, Stefan highlighted that "[y]ou don't talk with parents about this; you don't talk with your teacher, but at least you trust the school nurse", and Johan said that they realized that the school nurse is "a person that the girls feel confidence in [and] that [therefore] can be the one who will bring forward the message in the app". By Sara, Stefan and Johan, thus, the school nurse was stated as being a person girls trust and that girls feel that they can talk to. Making the app into a school nurse app was articulated as a way of communicating stable, correct information in a personal and trustful way.

How Sara, Johan and Stefan talked about trust in vaccination information implies that a distant and abstract way of communicating information would not be as efficient as letting a school nurse communicate it (that is, situating it). Distant and abstract views (facts) from nowhere would, according to my interviewees' reasoning, not be trusted in the same way as views from some*one*.

Using the figure of the school nurse as a person that generates trust since she has a close relationship to the girls has to do with the need to express care. Using the figure of a school nurse also includes a gendered trope of nursing as a caring profession. Enabling trust through closeness rather than distance and abstraction is about articulating a common assumption about care as being about closeness and warmth. The idea that a school nurse can respond to vaccination fears and worries through her way of being is linked to an imaginary of nursing as feelings-oriented care work. That is, it is not only a matter of close relationships, but about care *as* made in close, warm, and gendered relationships. Through the presence of the school nurse as presenting a caring view from someone, it is hoped that girls can be reached, and convinced.

The importance of receiving HPV vaccination communication from someone known is something that Leach and Fairhead discuss on the basis of conversations with parents about their vaccination decision. They write: "Encounters with professionals sometimes involve knowledge and information about vaccination – whether biomedical or not. Yet it is often less the knowledge dimensions of an encounter, than *the way relating to a professional* builds or undermines confidence, which shapes parental decisions about vaccination for their child" (Leach and Fairhead 2007:

76, emphasis added). In a related vein, Hobson-West (2003: 280, emphasis added) asserts that "[t]he policy of providing more information to the public [in her case, through risk-calculation oriented information leaflets] assumes that the information will be trusted [but] messages are *judged first by source and not by content*". Hence, these scholars highlight how trust in vaccination information depends on encounters and information source rather than content. This is a reminder of how Sara, Johan and Stefan stressed that facts communicated by a school nurse may more easily generate trust as they come *from someone*.

The vision of a view from someone includes an assumption about facts as stable and certain. That is, it is also a vision about *facts from someone*. When communicated by someone the girls trust and are close to, they will realize that the vaccination is safe and will trust the facts.

Conclusions: a clear-cut solution and its moments of friction

In this chapter, I have focused on how HPV vaccination information was discussed by the HPV app working group, and how their arguments were linked to the app and to how girls were thought to reason. I have discussed how my interviewees' articulations included three different (and related) dominant matters of care: *reaching girls where they are*, *good information as science-as-epidemiology* and *generating trust through a view from someone*. By conceptualizing these narratives and visions as matters of care (instead of stating that my interviewees are not doing care work), it has been possible for me to take seriously the specific articulations of care the interviews involved. For example, I have concentrated on the gendering of warm and close care in connection with the figure of the school nurse.

It also has enabled me to take seriously what my interviewees cared about in our conversations. As I have discussed, for them, myths, fears and rumors *are* worrying concerns, as according to them, these matters scare girls *unnecessarily*. Therefore, they strongly emphasized the need to do what they can to make girls less scared. Moreover, it has allowed me not only to discuss girls' fear, but also to take careful note of what the interviewees feared. For example, by focusing on their fear of the app being insufficient or failing and on their fear of uncertainty, I have tried to carefully attend to what this might say about the challenges of working in the midst of vaccination fears.

I have discussed how the app was invoked by some of my interviewees as a device that hopefully would reach girls where they are in life: as a care enabler and a good device for the job. *Reaching girls where they are* was a dominant matter of care. It was hoped that the app would make it easier for girls to get the information they needed to make a decision by bringing care to where they are. In this way, it was articulated as a matter of *mobile care* (in a quite literal sense!). The information was assumed as already there; it just needed to be transmitted to the girls. Thus, the app was articulated as a device for making already existing facts (which as some of them stressed may change over time) more accessible.

Reaching girls where they are is linked to another dominant matter of care that I discussed: *good information as science-as-epidemiology*. Good HPV vaccination information was envisioned as correct, impartial, stable and neutral, and as a clear-cut solution toward the problem with vaccination fears and rumors. It inhabited an idea that it would counteract such affective responses toward the vaccination, and therefore enable care for girls. It was hoped that science-as-epidemiology would abate feelings through facts, and that this would lead to a rationally based decision.

The third dominant matter of care had to do with *generating trust*. Lack of trust was envisioned as having to do with a lack of knowledge, and this was hoped to be solved through "good information". However, generating trust was also invoked as enabled through the presence of a caring school nurse that had a close relationship to the girls. Information was invoked as trustworthy if communicated from someone instead of merely in an abstract and impersonal manner. A *view from someone*, rather than a *view from nowhere*, was articulated as a way of enabling trust and care. Thus, the school nurse figured in the interviews as a care mediator and care enabler who could communicate facts in a caring and personal manner.

I have also problematized how these three dominant matters of care made other possible matters of care absent or marginalized. A predominant focus on side effects and the spread of rumors tended especially to make other possible things girls might fear, or be worried about, absent. Therefore, as an ethico-political commitment, I have allowed space-time for marginal and absent visions and narratives that were *already a part of* the otherwise often coherent narratives. In attending to these as moments of friction that rubbed up against the dominant matters of care, I have pointed toward glimpses of alternative narratives within otherwise dominant configurations. I have opened up glimpses of several moments of friction: facts as changing or as *uncertain certainties*, uncertainty – the "mini mini mini change" – as having a lingering presence, the precautionary action of stating *in essence*, and to hesitations, silences and laughter as interrupting the anticipated coherent meta-narratives. In different ways, these moments make present uncertainty in the midst of an articulated predominant need for stressing and confirming certainty.

In working with (often subtle) moments of friction, I have argued that slowing things down through attention to hesitations, silences and laughter can be to foster more livable matters of care. Concretely, it has made it possible to attend to how moments of friction can trouble and disrupt a trope assuming science-as-epidemiology to be a clear-cut solution to vaccination fears. Attention to these moments of friction, therefore, helps undo some of the tighter knots that present science-as-epidemiology as *the* solution. They make present certainty as something fleeting and contingent, and point toward worlds that are about "living with uncertainty" (Jerak-Zuiderent 2012: 736). Making these fleeting matters present, I argue, is about fostering glimpses of alternative, possible, worlds.

Whereas science-as-epidemiology as a guarantee for good information envisions care as an urgency (a solution is needed *now*), I have shown that slowing things down through attention to uncertainties made present through the interviewees' hesitations, laughter and silences has opened up for other matters of care. These alternative matters of care make it possible to approach vaccination communication as a matter of living with uncertainty. By attuning to these matters, it is possible to trouble a vision of a linear movement from uncertainty (feelings, stories, rumors, myths) to certainty (vaccination safety, factual information).

My attention to slowing down as a possibility for fostering alternative and more livable matters of care relates to discussions on *temporalities of care*. In contrast to how some feminist STS scholars argue for, or hint at, slowing down as enabling more careful and caring engagements (see Haraway 2008: 83; Jerak-Zuiderent 2013, 2015; Schrader 2015), Puig de la Bellacasa (2015) complicates such an idea. She argues that "care time" is "not so much about a slowing or redirection of timelines but an invitation to rearrange and rebalance the relations between a diversity of coexisting temporalities" (ibid.: 709). Her concern is that "advocating slowness [...] does not necessarily question the direction of the dominant timeline" (ibid.) Temporalities of care have coexisted in my discussion: a dominant narrative of the need for a clear-cut solution to vaccination fears, and more marginal ones about disruptive laughter, hesitations and silences, and about facts as for the moment stable (uncertain certainty), do, indeed, inhabit different temporalities.

The articulation that "good information" is needed to abate vaccination fears *now* presents action as a goal-oriented urgency. Additionally, to envision that more information "automatically" will transform affective responses (such as fear) into rational vaccination decisions includes a vision of a linear and goal-oriented temporality. In contrast to this, my chapter shows that uncertainties, disruptions and hesitations may open up for a multidirectional temporality of care. Thus, while articulations of a need for clear-cut solutions often include a goal-oriented, linear and "fast" temporality, slower temporalities (that coexist with the faster ones) may inhabit and enable uncertainties and hesitations that open up for alternative matters of care. This discussion of how slower and coexisting temporalities of care may disrupt and trouble calls for an urgency and necessity of vaccination will be developed further in Chapters 6 and 7.

EMPIRICAL PART II
CARING FOR COLLECTIVES, CARING FOR INDIVIDUALS

The first "I love me" campaign communicates the message that girls ought to "take care of themselves", and "take care of each other". Empirical Part II focuses on this campaign to discuss very different matters of care linked to individual girls, and to collectives of girls and others. It includes two analytical chapters: "Facebook Collectives Sparking Cares and Concerns" (Chapter 6) and "Girl-centered Care Trouble" (Chapter 7).

As mentioned in the methodology chapter (Chapter 3), the first "I love me" campaign was designed by Bredland County Council, and run between 2012 and 2013. It was designed before it was decided to increase the age span of the "target group", and was therefore directed toward girls between 13 and 20 years of age. As discussed in detail in the methodology chapter, the campaign consisted of diverse media and events (such as posters, the Facebook site and the "I love me" trailer tour). The campaign had a specific "I love me" logo that my interviewees explained depicts a heart that can also be viewed as a cervix (Figures 4 and 5 below). The campaign logo, and an extensive amount of the campaign's other materials, are designed in the color pink (Figures 4, 5 and 6).

Bredland County Council defined the campaign as a health promotion strategy that aims at *empowering* girls. In the campaign information can be read:

> The message and the mode of addressing should feel personal and desirous. The campaign should also be permeated with the idea of "empowerment", that is

> to strengthen the girls' own responsibility for their health and for getting vaccinated. And they get both practical and emotional tools for spreading the word: take care of yourself, you too! (Cited from campaign information from Bredland County Council 2012, my translation from Swedish.)

Empowerment is a notion that today is an integral part of different areas such as, for example, public engagement policies, girl-centered "girl power" initiatives and public health interventions – and is seen by many as perhaps the most central feature of health promotion work (Lupton 1995; Korp 2004). A common way of conceptualizing *individual* empowerment in a public health context is similar to how it is phrased in the first "I love me" campaign: it ought to enable individuals to take control over their health, to foster independence, to make them capable of taking responsibility for their health and making choices concerning their lives and health (see Nutbeam 2000).

The first of the two chapters of Empirical Part II focuses on the Facebook campaign site. The aim with this site was that it was going to be an arena for discussions about young girls' health and to make it easy for involved parties to publish comments, and like and spread the "I love me" message (campaign information from Bredland County Council 2012). In this way, together with the overall aim of the campaign as an initiative to encourage girl empowerment, the goal with the Facebook site was reminiscent of how, as discussed in Chapter 1, social media are currently embraced by health promoters as a promising new, participatory arena for public engagement (Korda and Itani 2013), patient empowerment (Neiger et al. 2012), and for reaching teenagers in their own arena (Ralph et al. 2011).

Chapter 6 discusses how different publics (for instance girls, parents and vaccination critics) used Facebook devices (such as the commenting function and the like social button) to confirm or critique the "I love me" message, and to voice matters of care and concern. I analyze how devices, publics and the county council on the "I love me" Facebook site articulated often conflicting versions of care. I explore how diverse actors troubled the empowerment message of "I love me", and articulated alternative matters of care. In particular, I discuss how actors articulated a critique toward an "I love me" vision stating that action needs to be taken *now* ("get vaccinated now!").

In striking contrast to the analysis of the Facebook site, in Chapter 7 I attend to how my interviewees emphasized how the first "I love me" campaign was meant to empower girls to take care of themselves. I dis-

Figures 4–5. Campaign images from the first "I love me" campaign. "Take care of yourself this summer! Get vaccinated for free against cervical cancer now".

cuss how my interviewees articulated the first "I love me" campaign as a matter of care *since* it was about girl empowerment. However, the chapter also focuses on how my interviewees themselves complicated and troubled *how* the empowerment message was formulated in the campaign. Following on the discussion in Chapter 6 on actions troubling an anticipatory message (that action is needed *now*), the chapter emphasizes moments in the interviews that complicated such a call for anticipatory immediacy, and that also made present other temporalities of care.

Figure 6. Campaign image from the first "I love me" campaign that depicts the "I love me" trailer. On the trailer is written "Get vaccinated against cervical cancer now!"

The two chapters in Empirical Part II deepen a discussion around temporalities of care. They center on questions of how individuals and collectives get linked to, and articulate, very different temporal matters of care. In so doing, the two chapters complicate promissory visions of "I love me" and the need for immediate action, and engage in a discussion concerning the ethico-politics of alternative temporalities of care.

6. Facebook Collectives Sparking Cares and Concerns

In the two chapters on the HPV app, I discussed how the app and interview conversations about the app focused on "reaching the girl" and "good information" as matters of care. As will be discussed in the next chapter, such a prevalent focus on girls was also invoked in the interviews about the first "I love me" campaign. In this chapter, I will engage in a very different story by analyzing re-presentations, mediations and articulations in a setting of the "I love me" Facebook campaign site. On the "I love me" Facebook site, actors were diverse: different human subjects and Facebook devices together often sparked conflicting visions of what care was. I pay particular attention to the entanglements of publics, the county council and digital devices, and how processes of liking, sharing and commenting generated matters of care. Attending to these matters, I focus on care specificities made in material-semiotic processes of digital mediation. Thus, the chapter illuminates how care is done on the Facebook site, and what such doings of care may say about the promises and troubles of care.

In the chapter, I combine attention to matters of care with attention to public participation in digital media. As STS scholar Noortje Marres (2005, 2007) argues, issues spark publics into being; without an issue, there is no public. Drawing upon, yet reworking, philosopher John Dewey's pragmatist theory of the public, Marres attends to public participation as a socio-material "practice that is occasioned by issues and dedicated to their articulation" (Marres 2007: 775). She explicitly discusses these articulations of issues in terms of care, as publics adopt "problems that no one is taking care of, so as to identify as an addressee for these issues that may take care of them" (Marres 2005: 216). Interestingly, similar to matters of care as an ethico-political commitment, Marres thus argues that by publicly caring for a *neglected* issue, publics are sparked into being. As Marres herself and others have argued with attention to the role of objects in public controversies, her approach is fruitful for studies of how social media devices participate in generating controversies in public matters (Birkbak 2013; Marres 2015). I will draw upon Marres's (2005, 2007) use of the notion of "issue publics" to discuss how the "I love me" Facebook site articulated matters of care, and sometimes matters of concern.

I approach public participation in this setting as involving, and enabled by, more-than-human (care) collectives. However, as I will discuss, not all "publics" that populated the Facebook site were "issue publics"; not everyone cared for HPV vaccination or the campaign as "issues" in need of problematization and care. I also make use of what others have entitled a "device perspective" to social media, something that enables a focus on how Facebook devices are participants in more-than-human (care) collectives (see Van Dijck 2011, 2013; Gerlitz and Helmond 2013; Ruppert et al. 2013; Gerlitz and Lury 2014; Weltevrede et al. 2014). Thus, the county council, publics and Facebook devices participated in assembling different matters of care and matters of concern. As I will show, different devices helped to spark a controversy over whether the "I love me" campaign and HPV vaccination were matters of care, and if so, how. In other words, devices enabled publics to, as STS scholar Mike Michael (2012) calls it, "overspill" the county council's framing of the Facebook site as an arena for *specific* (HPV vaccination promoting) public engagements. For example, by making present uncertainties and complexities, a vaccination-critical public together with Facebook devices articulated that, instead of being about caring for girls, HPV vaccination and the campaign were generating harmful effects, and that it therefore was a worrying matter of concern in need of public attention.

I am not the first STS scholar to discuss public participation through a matters-of-care lens. Both the promising (Pérez-Bustos 2014; Marks and Russel 2015; Felt 2016) and more troubling dimensions (Viseu 2015) of care practices have been discussed as part of public participation or engagement in technoscience. However, while these discussions attend to the involvement of STS scholars and lay people in the conduct of natural sciences, I discuss how digital media devices and publics participate in sparking different matters of care. My approach is akin to the one put forward by feminist technoscience scholars Kristina Lindström and Åsa Ståhl (2014) in their PhD thesis *Patchworking Publics-in-the-Making: Design, Media and Public Engagement.* Like they, I am interested in how digitally mediated "publics-in-the-making" enable matters of care, *and* how care as an ethico-political commitment can help foster an understanding for how, in this setting, "care *for* also means to see oneself as part of an issue and thereby also being partially responsible" (Lindström and Ståhl 2014: 312, emphasis in original).

I attend to the "I love me" campaign as one inhabiting a message of love as care. I am interested in the "promises and pitfalls of love as an affective political tool" (Myong and Bissenbakker 2016: 129) and how this on the "I love me" Facebook site includes specificities as to how care as an affective phenomenon is done. Importantly, I attend to the stakes in how love is re-presented and mediated and how this relates to a presence of other feelings such as happiness, fear and anger. How does Facebook mediate the "I love me" message of love (as care)? How is it possible to understand a presence of happiness, fear and anger in such a context? In focusing on care as both happy feelings and unhappy feelings on the "I love me" Facebook site, I emphasize the importance of focusing not only on the positive aspects of care, but also on "care's darker side" (Martin et al. 2015: 627), including *the politics* of happy feelings. I will show that the "I love me" message on the "I love me" Facebook site was simultaneously embraced by actors as a promise for happy and healthy futures and as a trouble in need of problematization.

I start with a discussion about some of the campaign images uploaded onto Facebook, followed by one about how Facebook devices sparked very different matters of care and sometimes matters of concern. The analysis of the campaign images shows how care is re-presented in the visual campaign material, and will illuminate an "I love me" vision (promise) of care as love (and happiness). Moving from there to how people actually related to this campaign on Facebook, and made use of Facebook devices in doing so, I show how other matters of care were engaged, generated and debated. I end with a discussion of what this case says about the stakes in Facebook-mediated care in a setting of health campaigns, and in turn, what this Facebook site can say about care.

Care as a pink promise about happiness and love

In one campaign image posted on the Facebook site (Figure 4), three girls are depicted together. They wear "I love me" bags with the "I love me" campaign "pink heart" logo. The three girls have hairstyles that accord with prevalent gender norms; they wear their hair long or in a bun. They walk arm-in-arm and they look happy. It is a photo depicting an everyday situation of friends hanging out in town. The girls perhaps have a free period from school. The image encourages the viewer to associate its message with youth and how it is to be a young girl.

Next to the image of the girls, the following phrase is written in pink: "Vaccination against cervical cancer free of charge. Bring a friend and do it today!" It was posted on Facebook by the county council together with the following status update text: "Do you have a hard time getting away to take the cervical cancer shot? Bring a friend so that you can support each other!" The "I love me" logo serves to present it as a message about love. Using the wording "friendship", the image tells us that the girls imaged are friends, and not, for example, sexual partners. It presents a narrative about self-love and friendship-love.

In three other campaign images posted on Facebook similar visual and textual narratives are told. In one of them (Figure 5), a girl is sitting on a bench, dressed in shorts and a white tank top – typical summer clothes. On her left she has placed the "I love me" bag. In another image, two girls are lying on the grass, both wearing summer clothes. There is an "I love me" bag next to them. Both images are accompanied by the phrases, "take care of yourself this summer!" and "get vaccinated against cervical cancer now", written in pink. In a third example, the same image as the first one is accompanied by the text "Nothing is more important than you!", also in pink. Along with the image is the following status update: "Show that you care about yourself and your body – get vaccinated against cervical cancer now! […] Spread and share with your friends!" Printed together with the "I love me" symbol of the pink heart, the message is one of both care for the self and care for others *as love*. Importantly, to share here means to use the share social button on Facebook.

In all four images, care is displayed as something you can do for yourself: take care of yourself and your body! At the same time, with the focus on friendship, another version of care is articulated. The encouragement to bring a friend so that you can support each other denotes care for others in terms of care for friends. Thus, two different cares are articulated: self-care and care for friends.

In these four images, Gardasil is re-presented as a vaccine against cervical cancer, something that makes absent the quite complex link between HPV infection and cervical cancer. Instead of presenting these complexities, a straightforward link between the vaccines as preventative technologies and cervical cancer is presented. Quite strikingly, there is nothing in the images that speaks of HPV, sexual infections, sexuality or even risks. Instead we are confronted with happiness, health, youth and a positive ambiance.

As feminist anthropologist and STS scholar S. Lochlann Jain (2013) discusses, it is common in cancer related health promotion campaigns to make absent cancer as pain, anxiety and possible death, in favor of a focus on happiness, the positive and healthiness – something that she argues is a good example of how cancer in our contemporary society is everywhere, but at the same time nowhere. In the four images it is striking how caring for HPV vaccination and cervical cancer is about happiness and love rather than about pain, fear and vulnerability. Care is made present in order to get a positive spin on the HPV vaccination message and to promote HPV vaccination. This is a version of care that equates it with positive feelings of happiness and love. As already mentioned in Chapter 2, equating care with happy or positive feelings is critiqued by Murphy (2015) as it presents care as something innocent and always for the better. She shows that this makes absent the politics of care, such as how care may assign responsibility to subjects and articulate a gendering of individual empowerment.

The four images on Facebook are good examples of a gendering of individual empowerment, in how they make present positive feelings of happiness and love as a matter of self-care and friendship-care. By invoking normative girlhood, they link love and care in a way similar to how feminist scholars Lene Myong and Mons Bissenbakker (2016: 134), in their discussion of a campaign using love as an affective tool, argue that "love can be said to establish a common orientation towards an ideal". Being oriented toward normative girlhood, this version of care as love comes with assumptions about the right way to care, including assumptions about what to care for, and who are the subjects that ought to be cared for. It locates love (and happiness) within certain subjects and groups and therefore promotes some lives and not others. As such, the happy feelings re-presented in the message of "I love me" come with exclusions.

In the images, there is an emphasis that girls should "get vaccinated now!" Thus, time is a part of what is happening here. The message is that sooner is better than later. Getting vaccinated is something that you ought to do *now* to prevent something from happening later on, in the *future*. This articulation of time displays a temporality of care: it is something that you need to *have done*. Thus it is not articulated as a continuous process. It troubles linear time: the future is brought into the present and ought to be acted upon now.

Puig de la Bellacasa (2015) critiques a linkage between care and time that asks us to take action *now* ("get vaccinated now!") as this relies on a politics of anticipatory immediacy. This is also very apparent in my material. As discussed in Chapter 2, such anticipatory politics often articulates a necessity and urgency of action, and makes absent complexities, contingences and uncertainties.

In the four "I love me" campaign images, such anticipatory immediacy is made possible through a reduction of complexity regarding linkages between HPV and cervical cancer. The message that girls *should* get vaccinated now is made possible through the absence in the images of non-gender-specific and non-unidirectional links between HPV and cervical cancer. This, together with the presence of happy feelings of happiness, self-love and friendship-love, makes a message about self-care as something to do now possible. Thus, the "I love me" message is made possible by a linking of desires for normative girlhood with an anticipatory immediacy that connects care and time in a progressive and urgent manner.

As already partly discussed, the message about love and happiness as self-care and friend-care comes with inclusions and exclusions. Some forms of love and happiness (and some subjects) are made possible and others impossible. Quite strikingly, these images invoke love as a "cluster of promises" (Berlant 2010: 93) that girls need to align themselves with in order to identify with the "I love me" message. To identify with the message, they need to align themselves with a version of love that is directed toward the self and to friends, that is closely linked to happiness and normative youth and that comes with a gendered call for caring for cancer through a "pink-washing" of cancer. As inhabiting a "cluster of promises" the "I love me" message promises – if you act now – a love that will make possible a healthy, normative and happy future.

Sharing a message about care as happiness and love

Encouragement to share the message as part of the "I love me" Facebook was articulated extensively by the county council and re-presented through campaign images uploaded on the site. For example, the encouragements "spread and share with your friends!" and "Who do you care about?" followed by "Share 'I love me' and take part in saving lives" were uploaded together with campaign images. In these examples, a message about self-love is being connected to a message about care for others in how people are asked to save lives by sharing the message on Facebook. Sharing is here invoked as caring; *sharing is caring* is the message.

The encouragement that sharing is caring and that people therefore should share the message with friends means, on Facebook, to share a message about HPV vaccination as healthiness, love, happiness and care. As Signe Rousseau shows in her (2015) chapter "Is Sharing Caring? Social Media and Discourse of Healthful Eating", such a message is strongly linked to normative ideals for what counts as a healthy and happy subject. Similarly to this, in the campaign images discussed above, the gendered articulation that girls should get vaccinated now as an act of care is made possible through sharing and liking. An encouragement to girls to act now gets translated into encouragements to click on the like and share social buttons.

The "I love me" campaign vision includes an idea of that Facebook promises an experience of the now, and that the platform promotes and mediates immediacy: care now, like now. This is reminiscent of how Kember and Zylinska (2012: 163, emphasis in original) discuss how Facebook promises a "*feel* of now now now". On the "I love me" Facebook site, care as anticipatory immediacy becomes re-presented through promises of Facebook devices as enabling immediate care.

The "I love me" campaign is envisioned to make it possible for people to align themselves with the "I love me" campaign message through liking and sharing as a confirmation or affirmation. It is envisioned that they like and share because they *do* like the message and they *do* care for, and love, themselves and their friends, and therefore want to share the message. This is helpful for the "I love" Facebook message; it can be further strengthened by social buttons. When people click on the like button, specific versions of care and love are affirmed and turned into a number. In this way, when the county council encouraged people to like and share the message on Facebook they also *asked* people to align themselves with the happy feelings message of "I love me".

Worth pointing out, however, is that even if people clicked, for example, on the like button on the "I love me" site does not mean that they automatically *did* like the status updates and comments. Social buttons can be used in ambiguous ways.

To promise quantified care

The campaign images and videos discussed above were liked on Facebook, and some of them were shared. For example, the image of the girls walking arm-in-arm was liked five times and the "Nothing is more important than you!" image was liked eight times when uploaded on Facebook. Comments on these status updates were also liked. The above mentioned status update, including the "Who do you care about?" image, was liked four times but not shared once. It seems that numbers matter here; if so, how?

That numbers mattered was re-presented in the campaign images. For instance, the "Who do you care about?" image was posted together with a comment from the county council. "Now we have 1,473 followers here on I love [me] and we need your help to become even more! Do all your friends know about the important message about free vaccination?" This message includes an explicit encouragement from the county council to like the "I love me" site and increase the numbers of followers. A similar example is a status update with an image depicting the words "Nothing is more important than you!" – an image that was posted together with the following message from the county council: "Now we have 1,929 likers here on I love me! Be a part of spreading the important message and help us to become even more! Share I love me to someone you like today!". Yet another example is an image depicting the "I love me" symbol with the heart. One status update including this image was posted together with the message: "Now we are 1,400 that like I love me on Facebook, tell your friend and share the site so that we can become even more!"

These Facebook updates re-present numbers of followers (hence, the number of followers that have clicked like on the "I love me" Facebook site) as indicators for how well the "I love me" message is communicated. The message to the publics concerned was that they care if they helped to increase the amount of followers on the "I love me" Facebook site by using the social buttons. On the site, therefore, a message about care was articulated as a message about liking and sharing as a means of showing that you care. According to this message, more likes and shares equals more care.

Facebook was thus drawn upon to articulate a quantification of care. Care became valued based on the number of followers, that is on the numbers of clicks on like. It became valued based on an *amount of* care, counted based on amount of Facebook device generated numbers. In the status updates mentioned above, 1,473, 1,400 and 1,929 followers were invoked as good care but *not good enough*. As these numbers were articulated as *not enough care*, the county council could encourage people to share and like to generate *further care*.

In this vision of the "I love me" site, numbers matter in a specific way. In a vein related to how STS scholar Helen Verran (2010, 2011) argues for numbers as material-semiotic devices that participate in processes of ordering and valuing, the numbers of followers re-presented in these status updates by the county councils were drawn upon to articulate care as something valuable.

According to a "quantified care" version of care, the "Nothing is more important than you" image generated most care as it generated eight likes, while the other only generated four and five. As I will discuss, some other images uploaded on the site as status updates by the county council generated hundreds of likes. Furthermore, many of the critical comments generated considerably more likes than other comments. This is a good example of how "numbers [on Facebook] have performative and productive capacities [as] they can generate user affects, enact more activities and thus multiply themselves" (Gerlitz and Helmond 2013: 1360). However, as I will discuss, *how* numbers mattered when they met different publics was not simply in line with how the county council re-presented it through the status updates about numbers of followers, or through the campaign images. That is, the vision of quantified care as based on the amount of followers, or as being about liking the "I love me" message, did not fully match with how differently publics and Facebook devices together articulated numbers as part of the doing of matters of care and matters of concern.

Disputing HPV vaccination through commenting and numbering

As brought up in the introduction to this empirical part of the study (Empirical Part II), the idea behind the "I love me" Facebook site was that girls could ask questions and get answers from the county council or from each other. The county council envisioned these questions as being about, for example, risks with getting vaccinated as well about where and how to vaccinate.

Such questions were also posed by girls and young women. For example, when the Bredland County Council posted an update with the information that the vaccination was now given free to girls and young women up to age 26, some girls and young women wrote and asked about how, and where, they could get vaccinated. They wrote things such as "Is this only available in [Bredland]?"; "Is it possible to get vaccinated if one is pregnant?"; "How many shots should one take?" These comments from girls and young women were in line with the county council's vision since they were concerned with matters such as how and where to get vaccinated. The questions were posed so that a clear answer could be written by the county council, or by other girls or young women. The county council answered these questions by articulating information and facts, and through encouraging statements: "Yes, so far it is only in [Bredland] county council that young women up to 26 years old can get free vaccination"; and "We absolutely think you should take the first shot as soon as you can [before you turn 26]!".

The questions posed by the girls and young women did not only generate answers from the county councils, or comments from other girls and young women wondering about similar things. They also provoked comments from vaccination critics warning the Facebook users who stated that they were thinking about getting vaccinated, saying for example, "Don't blindly trust all marketing! Investigate this further before you get vaccinated!!!" as one user wrote. Another wrote:

> I hope you who are writing that you think about getting vaccinated really read the declaration of content telling you about the injuries and side effects this can generate. [...] Be careful about your only life and get informed! That pregnant women even think about getting vaccinated shows how ignorant some are... [It's] horrifying.

In these two comments, the girls and young women who posed questions to the county council were encouraged to be informed before they decide to get vaccinated. Moreover, in encouraging people to be careful about their lives, the second user indicated that getting vaccinated is not being so. The commenters trouble the "I love me" vision of the urgency of HPV vaccination ("get vaccinated now!"), by encouraging people to learn more and wait before making a decision. These users strongly distance themselves from the county council's message, and through the commenting device articulate a different matter of care than the "I love me" version. Instead of vaccination being a way of taking care of yourself and your body – and as something to do *now* – it is by them articulated as something that likely will harm you, that is not about self-care and therefore needs to be carefully considered. These two matters of care come with different temporalities: one inhabiting urgency ("get vaccinated now!"), the other a slowing down of the "I love me" vision of anticipatory immediacy (think and be informed before you act).

The two messages from vaccination critics were the ones that generated the most likes as part of this commenting thread (under the status update from the county council discussed above): that is, 22 and 16. In contrast, the comments from the girls and young women and the responses from the county council discussed above received only one or two – and even zero – likes. However, the status update from the county council generated 409 likes (which in total was the highest amount of likes during the time the site was running). In addition, it received 246 shares. I will return to how numbers matter as material-semiotic devices in this situation.

Another status update from the county council stated, "Research shows that the vaccine against cervical cancer and genital warts doesn't generate any severe side effects". This comment generated 101 likes, 32 comments and one share. Many comments were critical: "Funny joke! Research that is only financed by pharmaceutical companies. You don't fool anyone but yourself!" an user exclaimed in the first comment posted as part of the commenting thread. This generated five likes. The second comment was from an user who replied and wrote, "that people are so anxious about side effects, [it's] embarrassing, girls!" This was followed by a comment from a girl stating that the "I love me" campaign is "awesome", followed by two hearts. None of these comments received any likes. Thereafter, the first user wrote back to the user who wrote that it was embarrassing that people were so anxious about side effects, and exclaimed, "You are surrounded by ignorance!" This comment generated three likes.

The comment was followed by a response from the county council where they referred to a movie from the National Board of Health showing "how a pharmaceutical becomes approved in Sweden". In addition, it was followed by a response from the user who was accused of ignorance. "And you as well, dear Maja. Don't believe in everything you read online", she replied. This comment received zero likes. The first user, in turn, responded by referring to a vaccination-critical site "Mothers Against Gardasil". This generated eight likes and was followed by a new user (Sofia) entering the scene, who responded to the critique put forward. "Maja, you need to be a bit more critical toward what you read online. Who has written the site? For what purpose?", asked Sofia. The comment generated zero likes. Again, Maja responded:

> [Y]our previous posts show an ignorance regarding the debated issue. A remarkable number of studies across the world have been conducted to examine Gardasil and its visible consequences from a health perspective. In several scientific studies, it has been shown that Gardail does not prevent HPV.

This comment generated two likes, and was followed by a reply to Maja by Sofia. She wrote:

> Please, post a link to the studies you write about instead of linking to unreliable homepages without authors. Gardasil has fantastic visible effects from a health perspective! I'm completely convinced about that, and I have enough references so you don't need to link to that . […] I really don't understand from where this unmotivated, completely disp☺portionate resistance comes from.

This comment received two likes. It was followed by several comments similar to the ones already brought up. In the commenting thread as a whole, people disputed whether Gardasil was safe or not, and whether the side effects were severe or not. These comments on the county council's status update are a good example of how side effects were strongly disputed on the campaign site through the commenting device. This is just one out of many examples where this happened.

Different publics did not agree upon whether HPV vaccination was safe or not, and the commenting device facilitated this dispute. Through the commenting device, people articulated strong opinions for *or* against. By using affective words and statements such as "ignorance", "completely convinced", "disproportionate resistance", "embarrassing" and "this is awesome!", through the commenting device they invoked their own perspective as the right one. Furthermore, in combining this with different forms of evidential claims, they blended affective registers (embarrassing, awesome, etc.) with factual registers (references to studies, etc.). For example, whereas Maja was stating that several scientific studies show that Gardasil does not prevent HPV, and that pharmaceutical companies distort the scientific accuracy of HPV vaccines, Sofia disputed this by problematizing how Maja was not including references to these studies and by implying that Maja was one of the users who articulated "disproportionate resistance".

Similarly to the last example, with the status updates including the information that HPV vaccination is now provided free to girls and young women up to age 26, the critical comments mentioned above were liked more times than the positive ones in favor of vaccination. All but one of the latter received zero likes (one received two). As I have already emphasized, even if it is impossible to know that all these users in fact clicked on the like social button because they liked the critical comment, the number of likes on the critical comments presents a contrast between the county council's vision about how Facebook would facilitate HPV vaccination communication (such as a quantification of care aligned with the message of care as love and happiness) and how it unfolded in practice. In general, on the Facebook site, the majority of the critical comments received more likes than the ones in line with the county council's vision.

At the same time, some of the status updates from the county council that presented a lack of severe side effects or new information about who is allowed to get vaccinated for free, received over 100 likes. Moreover, one of the status updates was shared 246 times. Also, many other updates from the county council were frequently shared. So it is not simply so that the critical comments were the ones that received the most likes. However, that some of the county council's status updates were shared and liked extensively suggests that devices were used differently by vaccination critics and people in favor of vaccination. It seems that many positive users did not engage in the conversation on the "I love me" site using the commenting device. Likely many users sometimes clicked like on status updates (and followed the site) and sometimes shared a message to enable more friends to get vaccinated, but perhaps they did not do more than that. This denotes a difference in how Facebook devices seem to have been enrolled: while critique seems to mainly to have been mediated by the commenting device together with the like social button, non-critique of HPV vaccination and "I love me" seems mostly to have been mediated by the like button and the share button, but not the commenting one.

That comments critical of vaccination in general were liked more extensively than positive comments demonstrates how the like button participates in valuing HPV vaccination and the "I love me" campaign. Returning to the vision of the "I love me" campaign, where the like button was supposed to be used to affirm the message about care as love and happiness, how it turned out was more complex and diverse than that. Contrasting versions of quantified care were articulated: both as shares through status updates and as vaccination critique through likes and commenting. This means that the like button, when aligned with a specific public, staged vaccination critique as more valuable than vaccination promotion. In doing so, it participated in articulating a matter of care that problematized the happy feelings of the "I love me" message through numbering practices. More quantified care according to this vaccination-critical version had to do with encouraging people to get informed and slow down their decision, as well as by emphasizing alternative evidence (e.g. critical studies).

As I have shown, status updates from the county council and comments from girls asking questions about the vaccination provoked responses from an upset and angry public. It is interesting that neither the county council nor Facebook were able to divert people from emphasizing critique and anger rather than happiness and love about, and for, HPV vaccination. The like button did not only affirm the "I love me" message. When the county council and Facebook met a public critical of vaccination, actions took a different route. Both the county council and the people expressing themselves as positive toward the "I love me" message and HPV vaccination had to argue *for* vaccination by positioning themselves as *against* vaccination critique.

The actions from vaccination critics sparked different matters of care and concern. While one public articulated HPV vaccination as a matter of care through the commenting device, another used the same device to articulate HPV vaccination as a matter of harm, worry and concern. Getting informed was staged as a matter of care. What it means to take care of one's life was in dispute: is it by getting vaccinated, or by getting informed about side effects and risks before making a decision? This dispute shows that the county council's vision of quantified care took another route. Care as in line with the "I love me" campaign vision of happiness and love was not simply facilitated by the like and commenting devices; instead it was disputed whether this was a matter of care at all.

I will now move on to discuss the dispute on the Facebook site about matters of care and concern in more detail by further elaborating how happy and unhappy feelings were articulated, and how this complicated the "I love me" vision about happiness and love.

"Hurray, kill the cancer!!!!!" – care as happy feelings

> We have noticed that several people have written negative comments about the HPV vaccine. This we feel is a shame as it protects against the virus that causes 70 percent of all cervical cancer. It's a serious disease and many are inflicted… […].

This status update was written by the county council in response to several critical comments on the Facebook site. By stating that several comments had been negative, the county council articulated this as a matter of concern, since the site was not being used to spread the positive message about care as love and happiness. This update generated a range of comments, comments that were both critical and appreciative. Similarly to many of the comments already discussed, they blended affective and factual registers to put forward their arguments. "It's so wonderful that people find protection against some diseases that threaten us", the first commenter stated. Envisioning cervical cancer as an external threat and HPV vaccination as scientific salvation, this commentator embraced HPV vaccination as a necessary cure against cervical cancer. Invoked as a vaccination against cervical cancer, an unidirectional and certain link between HPV, HPV vaccination and cervical cancer was imagined. In this way, the comment included a reduction of complexity. Here, HPV vaccination was invoked as a matter of care in how it helped people avoid disease. That is, it is made possible as a matter of care through reduction of complexity and uncertainty.

This was followed by other comments where people together with the commenting device articulated the "I love me" message and HPV vaccination as a desirable matter of care. For example, one user wrote: "Yeeeeeeees, who says no to medicine against cancer[?]". As a response to an user writing that she has gotten vaccinated and to a vaccination-critical commenter named Martin, the same commenter continued later on:

> U goo girl!!!! I can only say that I have taken the shot and I feel so good … I don't know, Martin, but are you a woman? All women are allowed to make this choice themselves. […] There is actually no one who wants cancer so no one will listen to you.

This Facebook user is arguing for HPV vaccination as a woman's choice and as a case of empowerment ("U goo girl!!!"). This is in line with the empowerment vision of the first "I love me" campaign. Moreover, she is using experiential knowledge in how she is emphasizing that she feels "so good" after having taken the HPV vaccination shot. Additionally, through the use of exclamation marks, her comment is formulated as an affective statement.

There is a tension here. At the same time as the user is stressing women's own choice she also states that no one wants cancer. Invoking the vaccination as a cure against cervical cancer and, simultaneously, as a "woman's choice", the vaccination is formulated as a good choice; as the right choice. If women want to take care of themselves, they should take the shots. In drawing upon gendered experience, the user invokes the idea that how one experiences and thinks about HPV vaccination is connected to one's gender. Women are presented as a homogenous category that consists of women who, on the basis of them being women with gender-specific experiences, are likely acting in favor of HPV vaccination.

This does not only say something about gender, it says something about Facebook and care. Even if the above commentator wrote a comment that was positive toward HPV vaccination and in line with the empowerment ethos of "I love me", she had to do so through a demarcation of her opinions and experiences as distant from the HPV vaccination critique circulating on the Facebook site. That is, HPV vaccination as a matter of care was articulated through a distancing from critique. Such distancing from critique is also evident in the following comment:

> How AMAZING that this is for women […] You women who are negative or suspicious and dare to [not get vaccinated], don't do it. I, on the contrary, who have had cervical cancer and have had surgery, I'm HAPPY and relieved that my daughter can protect herself from having to go through the same thing. So, she has already taken her two first shots. How could I as a mother and my daughter say no to that[?]. Hurray, kill the cancer!!!!!

In this comment, the commentator uses affective statements such as "AMAZING", "I'm HAPPY and relieved" and "Hurray, kill the cancer!!!!!" to emphasize, in a similar vein as the person writing the last comment, that HPV vaccination is a good cure against cervical cancer. Also reminiscent of the previous commentator, she uses experiential knowledge by drawing upon her own cancer diagnosis as a reason for why HPV vaccination is needed. Moreover, she distances herself from vaccination-critical comments on the "I love me" Facebook site. Like the last commentator, she draws upon a trope of choice when she states who people that are negative or suspicions do not have to get vaccinated. At the same time, she emphasizes HPV vaccination as a good choice when implying that she as a mother cannot say no to the vaccination. In doing so, she also draws upon a gendered trope of mothers' care responsibility for their children to enable an argument about the HPV vaccination as a good and needed choice.

There are many more examples on the "I love me" Facebook site of how happy affective statements and registers are invoked through a distancing from vaccination critique. For example, in the same comment field as already discussed, another user stated: "I am inflicted by cervical cancer. It is a given that I will vaccinate the daughter! Cancer sucks!!". Another example is an user who wrote "Finally!!! Damn, this is good!" after several critics voicing concerns. Similar to previous examples, these users formulated themselves in an affective and experiential manner to argue *for* HPV vaccination and *against* critique.

"IT'S ENOUGH NOW!!!!!!!!!!!" – care as unhappy feelings

As I have mentioned, several of the commenters from the above commenting field expressed critical comments toward the "I love me" message and HPV vaccination. Responding to one of the users expressing a positive statement about the "I love me" message, an user stated:

> Gardasil does not protect against cervical cancer … It protects against 4 out of 120 HPV viruses. And it is not at all clear that HPV viruses lead to cancer. Therefore, it is insane to pursue mass vaccination like it is done here. Tinni, what do you know about all the awful ingredients that are a part of the Gardasil vaccine?

In this comment, the user problematizes the idea that Gardasil protects against cervical cancer. In stating that Gardasil only protects against certain strands of HPV viruses, the user makes use of science-as-epidemiology to perform her argument. As a response to users claiming that Gardasil protects against cancer, she invokes complexity in how we can understand medical claims regarding cervical cancer, HPV and HPV vaccination. In contrast to the reduction of uncertainty evident in previous commentators' statements where Gardasil was articulated as a cure against cervical cancer, she brings up the non-unidirectional links between Gardasil and HPV viruses to argue against the need for HPV vaccination. Or rather, to argue that mass HPV vaccination is *insane*. Later on under the same comment field, the same user states "It is senseless that this continues. This false marketing should be reported to the police".

In using the wording "insane" and "senseless" this user invokes an affective register in a similar vein to how the people arguing for the vaccination used affective statements to put forward their argument. The difference is that whereas the previous examples included users invoking love and happiness to align themselves with the "I love me" message, this user invokes unhappy registers of indignation and worry to distance herself from it. Thus, the "I love me" message about love and happiness is actively resisted by this user through the use of other affective registers.

There are several other examples of users articulating that they are worried and upset. For instance, as a response to an image sent from the county council, one user wrote: "Fuck these [attempts to] frighten us that are supposed to make us run to the closest health care center to get the shoot. Tear this shit apart!" Writing in an affective manner, this commenter criticizes the "I love me" campaign for trying to scare people into getting vaccinated. Another example is an user who stated "IT'S ENOUGH NOW!!!!!!!!!!!!", implying that we need to stop vaccination as it hurts people. Yet another example:

> How can you so totally lack empathy, knowledge, intelligence and a heart! A heart that instead should be burning to protect our children and not the other way around! [...] You know that this is completely unnecessary, but, despite that fact, you endanger the health of so many people! [...] You're AWFUL!

As in the previous examples, these commentators use affective statements such as "IT'S ENOUGH NOW!!!!!!!!!!!!" and "You're AWFUL!" to critique the "I love me" campaign and HPV vaccination. In slight contrast to the commentator drawing upon a trope of science-as-epidemiology, this user makes use of affective registers without including scientific claims about side effects or complex links between HPV, cervical cancer and HPV vaccination.

In these examples where the commenting device facilitated vaccination critique, the "I love me" campaign and HPV vaccination were invoked as being the opposite of matters of care. Through the commenters' words, HPV vaccination and the "I love me" campaign were invoked as matters of concern in need of public attention as they harm the population. Instead these commentators articulated bringing up other different aspects of HPV vaccination, such as complexity regarding links between HPV, HPV vaccination and cervical cancer, as being about the commenters caring for the population, as they tell the public what is often hidden from them. This is possible through an idea of external, given facts as something the county council hides from the public. In stressing such an idea, these commenters articulated that the county council is choosing to do harm, knowing, as was implied in their comments, that the HPV vaccination is unnecessary and injuring the concerned population. In doing so, a delimitation is articulated between the county council as doing harm and critical people as caring through a making present of uncertainties. Here, HPV vaccination and the "I love me" campaign are seen as matters of concern in need of problematization and public attention rather than as matters of care.

The county council blocked some critical users on the Facebook site. When this happened, the county council posted comments about it on the site. One example:

> [U]nfortunately we have been forced to block some people from the site. As we write under the rubric "About" at the top of that page, under the ["I love me"] logo, we remove comments that are offensive, that include personal data which violates people's privacy, that can be understood as incitement to hatred or that for any other reason can be seen as inappropriate. For example, repeated comments with the same message or spam. The person who makes such comments risks getting blocked. We also remove comments with commercial messages.

As I will discuss further in Chapter 7, the county council eventually decided to close down the Facebook site due to the vast critique articulated on it (and because it took up too much of their time). This decision illuminates an asymmetry in how people can engage matters of care, and matters of concern. Being part of a campaign, the county council partly delimited what could or could not be said. Through closing down the site and blocking users, boundaries were drawn by the county council for what were the "right" matters of care and concern. Unlike Facebook groups established by lay people (see e.g. Birkbak 2013), the "I love me" group was initiated, and run by, the county council as an expert institution, and as such the county council could emphasize alignment with the "I love me" campaign vision as the "right" matter of care, and the vaccination critique as a matter of concern. In other words, the Facebook site was a setting of *public governance* of care, something that conditioned how matters of care and concern could be publicly articulated and engaged.

Device-mediated alignment and distancing

Both comments from users affirming the "I love me" message and from people critiquing it were liked and commented on. Facebook devices take part in generating and troubling matters of care and matters of concern. For example, when the vaccination-critical comments were liked, other users affirmed that HPV vaccination and the "I love me" campaign were troubling and harmful rather than matters of care. Simultaneously, through liking, they articulated vaccination and the "I love me" critique as a matter of care in how it made present complexity and uncertainty as well as the harmfulness of the HPV vaccination and the campaign. In a similar vein, people aligning themselves with the "I love me" message about love and happiness *as* care participated, through liking the comments, in articulating HPV vaccination and the campaign as a matter of the county council caring for girls and young women. In this way, Facebook devices are a part of different care collectives on Facebook, collectives that spark different versions of what care is.

This illuminates a generative potential of Facebook in how it can be many things for different users and that it, therefore, can be part of the perusing of very different politics (such as vaccination problematization and vaccination promotion). This is in line with how STS scholar Noortje Marres (2015) discusses a multivalence of Facebook that may enable a sparking of controversies or different concerns. Moreover, it shows how, on the Facebook site, publics were not only performing themselves in relation to the county council, but also in relation to other publics. Thus, digitally mediated acts of alignment and distancing – or identification and differentiation, as Michael (2009) calls it – relationally sparked publics into being.

Conclusions: devices and publics as care enablers and care troublers

I have discussed how versions of care were re-presented, articulated and mediated on the "I love me" Facebook site. I have especially discussed how the status update, like, share and commenting Facebook devices together with the county council and different publics articulated and sparked different matters of care, but also matters of concern. In doing so, I paid attention to how the "I love me" message about care, happiness and love was re-presented and articulated through different care collectives of images, devices, the county council and publics.

Different matters of care were re-presented and articulated. The campaign material images re-presented care as in line with the "I love me" message. In these, care became a matter of happiness, friendship-care and self-care, normative girlhood and a pink-washing of cancer. It also included a specific version of care as anticipatory immediacy, something that linked care to time through a vision of anticipation and futurity. As these images were posted on Facebook they also inhabited two visions of care articulated by the county council, and facilitated through Facebook devices: quantification of care and "sharing is caring". According to these visions, care is valued based on the amount of likes and shares. As part of these visions, anticipatory immediacy was articulated as desirable in how people were encouraged to like and share the message *now* to show that they cared about themselves and others. In this way, the like button was part of enabling a vision of quantified care in how they could be drawn upon as valuing devices that turned care into a numbering practice of anticipatory immediacy.

Numbering practices were not only in line with the county council's vision of "sharing is caring" and care as love and happiness. Even if the county council's status updates sometimes received the most likes and shares, vaccination-critical comments as part of the commenting device were in general liked more than questions to the county council from girls or vaccination-positive comments. Numbers participated in disputing HPV vaccination and the campaign as they took part in troubling the happy feelings message of "I love me". They staged critique as more (that is, higher numbers of likes) valuable than promotion. That is, the like button participated in a care collective which troubled that care is about the version of love and happiness articulated in the "I love me" campaign. Thus, quantified care was also articulated as a vaccination critique. This means that it was not as simple as that numbers were *either* facilitating vaccination promotion *or* critique. But different devices did different things during specific circumstances; whereas many shares and likes on status updates affirmed the message from the county council, many likes as part of the commenting device were mainly facilitating vaccination critique.

Through Facebook devices, the "I love me" site generated a controversy between vaccination promoters and vaccination critics around whether HPV vaccination and the "I love me" campaign should be understood at all as matters of care.

For a vaccination-critical public, HPV vaccination and the campaign were matters of concern in need of problematization and public attention. Through upset and angry affective statements, making present HPV vaccination uncertainties and complexities understood to be hidden by the county council was articulated as an important matter of care. Here, when HPV vaccination was articulated as a matter of concern, critiquing and "unmasking" HPV vaccination were presented as matters of care. This public enrolled the commenting and like devices as ways of pushing for a need for critique; vaccination critique became a matter of care through comments and likes.

A vaccination-positive public aligned themselves with the message of "I love me" to confirm that HPV vaccination is a matter of care, and did use the commenting and like devices for doing so. For this public, vaccination critique became a cause of worry that needed to be responded to.

Yet another public consisted of girls that sometimes posed questions and perhaps liked and followed the site and some of the status updates, but who did not participate in caring for HPV vaccination and the campaign as controversial issues. As such, these girls were involved in, and provoked, disputes on the site; but they did not actively engage in the conversation through comments or likes on comments.

The actions from these publics sparked a vaccination and campaign controversy where matters of care and matters of concern were debated. At the Facebook site, no closure around how HPV vaccination and the "I love me" campaign can be a matter of care or not was generated. Instead very different matters of care and matters of concern were sparked. I will end the chapter by discussing some of these, but first I will further discuss why it might be fruitful to use "matters of care" and "matters of concern" together.

In this chapter I have empirically used both matters of care and matters of concern to illustrate how these notions can be used together to point at multiple dimensions, or different sides, of the same phenomenon. As Puig de la Bellacasa (2012: 89) discusses, both can be understood as "affective states"; whereas care comes with connotations of "attachment and commitment", concern "denotes worries and thoughtfulness". And this, I argue, is precisely what I have shown was at play on the Facebook site. When the "I love me" campaign was articulated as a matter of care to align and commit oneself to through, for example, acts of liking and sharing the message, others responded by saying that the campaign and the vaccination were rather reasons for worry and thoughtfulness. Rather than depicting vaccination as a matter of urgency and immediacy, the critiques encouraged people to slow down and think before making a vaccination decision.

The Facebook site generates insights on temporalities of care. The vision of care as anticipatory immediacy encouraged people to care *now* through liking, sharing and commenting *now*. Different from this "I love me" vision, many critiques encouraged people to slow down to think about the decision carefully instead of rushing toward it. In this way, a coexistence of different temporalities of care was articulated on the site.

An important part that I have not yet talked about is how *the pace* of Facebook differed in relation to the commenting device. It was not only about anticipatory immediacy. If one status update was commented on by someone, this often quickly provoked a wide range of replies. During these times, the pace of Facebook was intense: proponents, critics and the county council responded to each other back and forth at a brisk pace. At other times, when status updates were not commented on immediately or quickly, they could remain without comments and likes for a long time, even for ever (many status updates, in fact, on the site remained without comments, likes or shares). During these times, the pace was slow, sometimes even motionless. Thus, different paces coexisted, or appeared and disappeared, during different times of the site's existence. In this way, Facebook devices were "pacing devices" (Weltevrede et al. 2014: 135) that opened up for temporalities other than immediacy, while also articulating immediacy. Moreover, they were what can be conceptualized as "pacing care devices". As such devices, they participated in generating different matters of care. Thus, paces of care coexisted during the existence of the Facebook site: some staged anticipation, whereas others slowed down HPV vaccination matters, and thus problematized their urgency.

In addition, the Facebook site provides insights about "public care" in technoscience as a meeting between public participation and care. Compared to "non-participatory" forms of health campaigns, when remediated into a campaign Facebook site, communication and engagement between publics and the county council were staged. In how Facebook opens up for dispute and conversations holds a promise for careful engagements, and co-learning, between experts and publics, and it holds a promise for inclusion of alternative visions and commitments, such as a "becoming with others" in a context of public engagements. As argued by STS scholar Ulrike Felt (2016: 193), it is important to attend to how public engagement events can inhabit a logic of care in which controversies or "detours" are seen as "valuable moments during which different perspectives are opened up".

However, and as will be discussed further in Chapter 7, "learning from publics" in a setting that is conditioned by asymmetries in power can easily reproduce moralizing assumptions about how such engagement should look, and about what visions and commitments are the "right" ones. Notably, and as the county council did, blocking users and closing the site down reaffirms certain versions of what is "good" care, and excludes others.

In general, at the same time as the Facebook site allowed for the inclusion of diverse publics to promote and trouble HPV vaccination (instead of only girls and the county council), how communication was re-presented included exclusions and moralizations, such as condemnation of other people's actions and opinions. Importantly, this goes for all sides. Both the county council and publics on the Facebook site articulated moralizations to present their issues and concerns. On the Facebook site, the focus from the different publics (not the girls posing questions) and the county council was mainly on what others *should* do, not what they *could* do. Others *should* get informed, *should* get vaccinated, *should* act now, and *should* like the message. Therefore, the Facebook site simultaneously opened up and closed down for difference. This highlights "public care" in public engagement with technoscience as both a promise and a trouble; as political, and as far from innocent.

The Facebook site generates insights into discussions on care devices. As simultaneously care enablers and what can be termed "care troublers", the different Facebook devices together with different human actors generated different matters of care, and troubled others. The devices participated in enabling and troubling different of versions of care in "dances of relating" (Haraway 2008: 25) that inhabited ongoing relational acts of distancing and alignment from different matters of care and matters of concern. For example, as numbering devices aligned with different humans, they were multivalent in how they participated in enabling a "sharing is caring" vision *and* troubling the same. That is, different quantified cares were facilitated and troubled by the devices. Therefore, in staying with the trouble, it is possible to see how devices (and publics) in different practices, and through different more-than-human collectives, can both enable and trouble care. I will further discuss the propositions of devices as "care troublers" and as "pacing care devices" in Chapter 7.

7. Girl-centered Care Trouble

We wanted it to be "as positive as possible", my interviewee the Head of Communications, Klara, stated when we talked about the empowerment message of the first "I love me" campaign. "It's an active choice to focus on the girls, to strengthen them" and it is "really a message about love", she continued. Emphasizing the focus of the girls as a positive message about love and care, Klara's words are reminiscent of my discussion around the campaign images in Chapter 6.

In the interviews, links, and sometimes tensions, between protection and empowerment, sexual activities and girlhood, girls as individuals and as a herd, HPV vaccination as anticipating the future or improving current life, as well as between facts and feelings, were articulated. All of these conjure articulations of how girls and care relate. How to care for individual girls? How to care for the herd of girls? How ought girls to care for themselves and others? Is it possible to care *with* girls? Accordingly, in this chapter, I discuss how girls and care were being linked in the interviews about the first "I love me" campaign.

Feminist STS scholars have pointed out problematic relations between girls and care in the realm of biomedicine. In a context of feminist STS takes on matters of care, Murphy (2015) illuminates troubling itineraries of how links between care, empowerment, and girls are configured. She shows that in the name of the girl, feminist strategies of empowerment can be co-opted to put forward the message that we *immediately* need to care for the girl, something that easily reproduces prevailing power dynamics. In a related vein, Roberts (2015) shows how links between care, girls and sexuality may articulate girlhood as a site for anticipatory politics that provoke, as she refers to it, intense feelings of concern, concerns that ask us to anticipate girls' futures. In a similar vein – and which is an explicit reference point for Roberts – Adams with colleagues (2009) show that girls are called upon in anticipatory regimes, something that provokes powerful, affective and temporal dimensions of care. Strikingly, these scholars use HPV vaccines as illuminative examples. Finally, scholars working on HPV vaccines, such as Laura Mamo with colleagues (2010), Jessica Polzer and Susan Knabe (2009, 2012) and Nichole Charles (2013, 2014), elucidate how discourses of empowerment, protection and care of girls may go hand in hand within a context of HPV vaccination campaigns.

Learning from these scholars, it seems that "the girl" is often articulated in ways that relate care (as empowerment *and* protection) to gender in problematic ways. What the above scholars perhaps first and foremost shed light on are the seemingly paradoxical ways in how practices of anticipatory immediacy may simultaneously include articulations of a need to care for girls, and of girls to care for themselves.

The concerns emphasized by these scholars serve as an important basis for this chapter. As I will show, some of these concerns are indeed staged in the interviews. However, as will be clear, other things are also at stake. Paying close attention to predominant, marginal and absent links between girls and care, my focus is, as in all other chapters, on the specificities of care. In taking an interest in articulations of different (predominant, marginal and absent) matters of care, I will pay close attention to the complex stakes of the girl-centered cares articulated in the interviews. I ask how girls figured in the interviews as subjects who ought to care, and that need to be cared for.

Some threads of this chapter will be reminiscent of discussions in previous, and upcoming, chapters. Similarly to Chapter 5, I will discuss tensions between, facts and feelings. Yet, if feelings as the basis for "good information" were predominantly troubled at Mittland County Council, in Bredland, feelings were sometimes embraced and sometimes problematized as a proper mode of address.

I draw upon interviews with the Head of Communications, Klara, the communicator Helena, the epidemiologist Emma and the administrator Linnea. As explained in the methodology chapter, they have all to different degrees been involved in work with the "I love me" campaign. I will begin with a discussion of articulations of the "I love me" campaign as a positive message about empowerment that aims to foster girls' care, love and responsibility for themselves and their health – and not about disease, infection control, death and sex (as risks). This will further deepen our understanding of the county council's vision of the first "I love me" campaign, introduced in Chapter 6. From there, I will move on to how the absence of any sexual dimension in the campaign was problematized, something that in the interviews complicated how the vision of "I love me" was configured. Following from this problematization, I will add more trouble by attending to how the communication on the "I love me" Facebook site was not invoked as a matter of concern in line with the "I love me" message. Partly echoing Chapter 6, this discussion will further unsettle the message of "I love me". Finally, I turn to how *listening to the girls* and *being where girls are* were articulated as fundamental to enable girl-centered, and therefore good, care. Attending to the specificities of these versions of care, I take an interest in the politics in how girls are called upon to be active participants of care. I discuss such girl-centered cares through attention to how different textual and methodological care devices were enrolled in the interviews: interview evaluations as care troublers re-presenting girls' opinions, and complicating the work of the county council, and the "I love me" trailer as materializing temporal, spatial and mobile care. In sum, this chapter explicates the stakes of girl-centered care, and deepens the discussion introduced in Chapter 6 on care devices for public participation in technoscience as care enablers and care troublers.

A message about girl empowerment

As indicated in the introduction to this chapter, the "I love me" campaign was described by several of my interviewees as being about encouraging and empowering girls to like, love and care for themselves. This directly relates to how the "I love me" message was discussed in connection with the visuals in Chapter 6, that is, in how they re-presented girl-focused care as happiness and love. For example, the communicator Helena said:

> The basis of our campaign is that we want to empower the girls. We want to, you know, get them to feel and think that "I do something good for myself. It's my own decision, but if I do this I think about my own health and I do it for myself". That is, you know, what's behind the message of "I love me".

The "I love me" campaign, in Helena's words, is about a message that getting vaccinated is something girls do for themselves and for their own health. In another interview, she said that the "'I love me' spirit" has to do with encouraging "a responsibility for oneself and one's health and that there's, you know, a decision to make about one's body and future". In a similar vein, the epidemiologist Emma emphasized that it is good that the focus of the campaign is on "*your* body, what *you* think of it, I love myself, I love me" and, as partly brought up in the introduction to this chapter, the Head of Communications, Klara, stated that it has to do with encouraging girls to "empower themselves" and that it is "a message about love". As such, for Klara, it is a "positive message" conveying a "positive feeling". She continued by connecting this to the "I love me" message about care: "[t]his 'take care of yourself' message is a commonly used positive message".

In these excerpts, the "I love me" campaign is stated as encouraging, and strengthening, girls' love, care and responsibility for their own health, body and future, and to make girls understand that it is their decision if they want to get vaccinated or not. The "I love me" message is emphasized as a positive message about self-care and individual girl empowerment.

This focus on girl empowerment comes as no surprise as it is widely circulating trope that often encourages girls to take responsibility for their health. This has been discussed in relation to HPV vaccination campaigns in Sweden (Lindén 2013b) and elsewhere (Mamo et al. 2010; Polzer and Knabe 2012; Charles 2013, 2014). In Sweden, it has also been analyzed as circulating in other contexts of girl-centered health promotion (see e.g. Oinas and Collander 2007; Söderberg 2011; Gunnarsson 2015).

In the interviews the focus on encouraging and empowering girls was sometimes connected to choice. When I asked Helena about the aim of the campaign, she answered:

> To give [girls] some confidence … That it's about their bodies and that they make a choice if they should get vaccinated or not … That it's about them caring for their bodies and lives. This since we believe it's a good thing. And the message is that if you have done it, then you have made a good choice. But it's up to the individual to make an own decision of course. You have to … If you see cons, then you have to weigh the pros against the cons.

Helena articulates that girls make a good choice if they get vaccinated. Yet, "choice" is also invoked as a "free choice" when she stresses that girls have to make up their own minds and balance the pros against the cons regarding the vaccination. Here, the girl who makes the decision herself is envisioned as a rational human subject that calculates pros and cons and, based on that, makes a good choice. So, this form of care as being about giving the girls some confidence comes with a vision of the girl as a calculative, rationally choosing subject.

The communicator Helena and Head of Communications Klara continuously repeated that the county council promotes vaccination since it believes it is something good. As Helena explicitly said earlier: "we believe this is a good thing". Elsewhere in the same interview she stated, "we do believe it is a good thing to get vaccinated of course". She emphasized that girls themselves ought to understand that it is something good and that the county council's task is to enable that understanding. A choice is not a good choice if it is not the girls who have made it *themselves*. Hence, the county council needs to communicate that vaccination is something good (it is their assignment to do so), but they cannot decide for the girls. They can only encourage, empower.

In a similar way, Klara emphasized that the county council needs to make sure that "everyone has access to good information about what ones' choices are". Hence, "good information" is what enables choice. Reminiscent of this, the county council administrator Linnea focused on young women's own choices as being important in how the county council should provide HPV vaccination information. She, too, stressed that to enable girls to choose they must be given "good information". She continued:

> "Do I want to get vaccinated or not?" That must come first. After that, the young woman needs to make the decision "should I go and get vaccinated?" And if they choose to do so, they must be treated in a good way. Or, perhaps, there are young women who do not have a lot of knowledge [...] [They] haven't thought about it, they don't know anything about HPV vaccination [...] And then you get the information then. Then you get a choice.

"Good information" is formulated by Linnea and Klara as information that enables choice. In Linnea's words, you get a choice when you have been given information about what HPV vaccination is about. Then you can make an *informed* choice. Providing good information that enables choice is invoked as an act of care that enables girls to make decisions on a well-informed basis. Girl empowerment was thus articulated as simultaneously about good, informed and free choices.

Troubling a care for the herd

In the above excerpts, there is a strong focus on the individual girl as the one who needs to be encouraged to decide, choose, and care. This focus was problematized by Emma. From the interview:

> EMMA: [...] As an epidemiologist, concerning vaccines is a lot about protecting others [...] This "I love me" [campaign] really has an ego focus.
>
> LISA: It's a lot of focus on the individual girl.

EMMA: Exactly. And perhaps it has to be like that. It's such a difficult balance between the social [aspects of] infections, as it's me who has the infection and that will be treated, I'm the one diseased. But from a social perspective, there is something more with infections, as you infect others and vice-versa. Thus, it is somewhat difficult when you only chose one of the parts and not the other. And now they have chosen this part [the focus on the individual].

Emma stresses that she thinks it can be a problem if one only focuses on individuals and does not include a social perspective on infections. A social perspective on infections denotes vaccination as something that can be done for the sake of the health of others; as a care for others. Intrigued by her problematization of the focus on the individual girl, I asked her to brainstorm how a campaign focusing on the collective and infection control in the context of HPV vaccination would look. She replied:

> It's difficult. You have the right to your body and you have the right to get vaccinated. You have the right to decide over your body. So I can also think that, in some ways, that you always respect someone's decision over her body [...] So, in some ways, now it perhaps sounds like I am switching from what I said earlier, but in some ways, then this campaign is good. That the focus is on *your* body, what *you* think of it, I love myself, I love me. It's, you know, about me. And I think that's completely right. That despite concerns for herd immunity and that you want to avoid unnecessary infections [and] spread of infection. But at the same time, you can never, I think personally, force vaccinations on people in any circumstances. Never, never, never. So it's maybe this self-focus that it needs to have.

Emma balances between problematizing the focus on individual girls and stating that it is the right one. It is striking that when the focus is on girls' own bodies, a right to choose argument "wins" over ideas for how a campaign could have included herd immunity and a care for others. This is reminiscent of Helena's focus on good choices as choices that girls make themselves. Emma, too, is focused on choices as something that must be made by individually choosing subjects. As is evident from the quote above, when I tried to push Emma to elaborate more regarding what a "care for others" campaign could look like, she seemed compelled to emphasize that it would be wrong to force anyone to get vaccinated. Hence, a care for others is implicitly turned into being about compulsion, something that is contrasted with choice. According to this reasoning, compulsion cannot be a matter of care, and as the other option is choice, choice "wins". A dichotomy is invoked: the freely choosing individual versus vaccination as compulsion. This brings forward a situation where the focus on "my body, my choice" makes "I care for others" impossible. Other possible ways to focus on the collective, or the herd, than compulsion and control are absent. That is, a care for the herd that is not about compulsion becomes unintelligible.

The herd was brought up, and troubled, in other interviews as well. For example, in the interview with Klara, the herd was invoked by an overview of my thesis that, based on her request, I had sent to her a few days before our meeting. As a textual device that organized, and transformed, our conversation, the thesis overview highly affected our conversation in how Klara returned to it throughout the interview. Klara continuously emphasized the importance of focusing on the goal of increasing the vaccination coverage as the basis for how the first "I love me" campaign was designed, and not, as she deduced from my thesis overview, the individual or the herd. According to her, my question about what information should be communicated did not make sense. It was not about the individual or the herd. A campaign's focus, she explained, depends on the assignment and the goal (that is, increasing vaccination coverage for this target group). Consequently, when the individual and the herd were evoked by my thesis overview, she stated the goal and assignment – and not the individual or the herd – as the important matter. From the interview:

> [W]hen we analyze what's effective in terms of communication, then we don't think the herd or the individual [...] Therefore, that question is not relevant for us in that way [...] I mean, we have this assignment to cover the whole target group and we need to reach a coverage of 95 percent. And then we find a concept [that aims to reach that goal].

Klara continued to emphasize that this means that they do not think in ideological terms about what is "right". Instead, a campaign idea that will enable the county council to reach the goal of increasing vaccination coverage for this group was stated as necessary. For her, this was the "positive message" of "I love me".

The idea that a focus on the individual girl would work, as it would make it possible to match the stated goal, is built upon assumptions about who the teenage girl is. The emphasis on the *goal* and *assignment* served to justify the focus on the individual girl and a "positive message" about love and self-care – and to justify the absence of a care for the herd in the campaign. This invokes a gendered trope of girls as empowerment subjects who are likely to align with an "I love me", "I care about myself" message.

Klara stated several times that her answer was related to her position as a communicator. "If I was a disease control doctor, then perhaps I would have thought differently about it", she said. In doing so, she implied that such a doctor would perhaps have focused on the herd rather than the individual. She, as a communicator, thinks about assignment and goal, but perhaps people working with disease control think specifically about the herd. This is reminiscent of how Emma made use of her position as an epidemiologist to problematize the focus on the individual girl. In both cases, a care for the herd is made present through a focus on the interviewees' professional identities, and it is done so *through* a focus on love and care for oneself as, in this situation, the "right" mode of address.

Troubling an absence of sexual dimensions

The majority of my interviewees did raise the fact that the county council decided to not include the sexual dimension of HPV vaccination in the campaign. For example, when Klara emphasized how the first campaign was about helping girls to "strengthen themselves", she then connected this to a wish to not focus on sexual behavior:

> KLARA: The basic idea during the first period [of the campaign was that] we didn't want to relate this to sexual behavior, but rather as something that was about the girls strengthening themselves. [...] And not that, you know, "here you put yourself at risk in life, now we want to help you to avoid that risk". But, instead, as something as positive as possible.
>
> LISA: So it wasn't a focus on risk factors and that kind of thing, then?
>
> KLARA: No [...] It's an active choice to focus on the girls, to strengthen them.

In this quote, helping the girls to strengthen themselves as a positive message is linked to a removal of sexual behaviors. Later in the interview she stressed this directly: "we focused on the positive [and we] removed sexual behavior". Thus, sexual behavior was contrasted with a positive message and was, therefore, evoked as something negative. This connection between sexuality and "the negative" has to do with Klara connecting it to risks, death and disease. They thought that in a context of HPV vaccination sexual behavior would become a question about risks, death and disease. This they wanted to avoid. I asked her to develop what she meant. She answered:

> KLARA: [...] Partly [it was due to] them being so young. And then it was also connected to [concerns we had about] connecting sexuality, which essentially is

something really positive, with disease and death. That, for us, was wrong [...] We also reasoned that these young girls often have to carry a burden, and a main responsibility, for [sexual] protection and other things. A big responsibility for sexuality.

LISA: Yeah, that's true.

KLARA: So, perhaps it was also time to lift that weight from their shoulders [...] And then really not connect it to yet another responsibility for disease and death [and risk].

In the campaign, Klara relates the decision not to connect the HPV vaccination to sexual behavior to two things: age and a female gendered responsibility for sex. The girls were young *and* the county council did not want to put yet another sexually related responsibility on them. Sexuality comes with specificities here. As already discussed, sexuality is by Klara expressed as *in the context of HPV vaccination* being linked to matters of disease, death and risk. But it is also stated as essentially something really positive. That is, *outside a context of HPV vaccination*, it can be something positive. The county council did not want to put responsibility for sexual behavior linked to risks of disease and death onto girls; it wanted to focus on positive aspects. As already discussed, in Klara's words "the positive" was about empowering and encouraging girls to take care of their lives, health and future; a responsibility to love oneself.

In the interview with Klara, a possibility for including sexual activities as something other than disease, death and risks, was not brought up. According to her line of thought, including a sexual dimension in the campaign had to be about "the negative" parts of sexual behavior and not about sex as "something positive". Thus, instead of, for example, using HPV vaccination campaigns as enabling an opportunity to discuss different dimensions of sex and sexuality, the answer becomes to remove the sexual dimension from the picture. Sexuality itself was not stated as the problem; it was rather that in connection with the HPV vaccination it *had* to be made present through negative matters such as risks. Protecting girls from *sexual activity as something negative* became a matter of care. However, and as I will discuss regarding sex in this context, what is negative and what is not so was not articulated as a given – neither by Klara, nor by the other interviewees.

The focus on sexual activity as risk, death and disease can be connected to discussions in girlhood and childhood studies where a "sex negativity" (sex as danger and risk) figuring in, for example, sex education, media and policy reports is problematized as picturing girls as innocent and in need of protection, as well as for enacting girls' sexuality as inherently problematic (see e.g. Egan and Hawkes 2008; Renold and Ringrose 2011, 2013; Sparrman 2014). This can be connected to Klara's emphasis on the fact that they did not want to include risk factors in the campaign; they did not want to include "sex negativity" as sex "essentially is something really positive". This differs partly from how in Chapter 4 I discussed how the app re-presents sexual activity as precisely a matter of risk (of "sex negativity"). In contrast, making the sexual dimension absent was through Klara's words invoked as a form of care *as* protection *from* gendered sexual politics.

In making absent the sexual dimension of HPV vaccination, care for girls as a protection from "sex negativity" taps into tropes that include assumptions about when, and how, girls should encounter sexual activities. Thus, to decide to not include a sexual dimension to lift the weight of gendered sexual politics from the shoulder of girls, is simultaneously to tap into assumptions of timing and appropriateness for sexual activity. Moreover, as childhood studies scholars Emma Renold and Jessica Ringrose argue, this relates to adult fears and anxieties about girls' sexual activities. These, they assert, tend to reduce "the complexities of girls' experiences of sexual pleasure and danger" (Renold and Ringrose 2011: 394). Thus, this view of care *as* protection from sex *as* risk closes down what teenage sexuality might be and become.

It is, however, a bit more complex than that. Klara stated several times that the decision to exclude a sexual dimension was "based on the knowledge we then had", and "we received data making it clear that this did not work". The "data" Klara referred to was an interview evaluation made by the communication division at the county council which, based on interviews with girls, articulated that even if girls liked the campaign, it did not motivate them enough to actually take the step to get vaccinated. Based on these findings, the county council came to a decision regarding a new campaign (the second "I love me" campaign, discussed in Chapters 8 and 9). Klara stated that, when listening to girls through the interview evaluation, new insights were gained. Klara explained that the evaluation showed that the girls thought it was necessary to include death and disease and, at least implicitly, connect these to sexual behavior.

Klara's answer invokes *listening to the girls* as a matter of care that complicated how they previously cared for girls through protection. This illuminates a contingency regarding what "the negative" and "the positive" are regarding sex in this context. Avoidance of a sexual dimension was problematized when this did not match the "data" from the girls in the interview evaluation. Caring for girls as in *speaking for them* was in tension with how the evaluation articulated that girls wanted to be cared for. In connection with this, Klara did not try to argue that the decision to exclude sexuality was necessarily a good one. She merely stated that this was based on what at the time they believed was the right way to reach the girls. But due to new data, this changed. Invoking data as situational and contingent, Klara emphasized the decision to exclude sexuality as not working. It was articulated as a *failing* matter of care.

The epidemiologist Emma emphasized that from the start she was critical toward the removal of a sexual dimension from the campaign. Based on results from an interview study she had conducted as an epidemiologist (not the same as the evaluation interview done by the county council) that, according to her, did show that girls care more about genital warts than cervical cancer, she told me that she tried to push for the inclusion of genital warts, and therefore sexual matters, in the campaign. I asked her whether the decision not to follow her suggestion was that people wanted to focus on cancer instead. She answered:

> Yes, on cancer. Exactly [...] But I believe that one could have made more girls to appear at the vaccination clinics [if genital warts had been included]. In this age group where one often is sexually active, if they [the girls] understood that "my God, I can, you know, get a protection against an STI [sexually transmitted infection], something that is so embarrassing, so awfully embarrassing" [...] It was so obvious [in the interview study she had conducted] that this was what was important for this group [as it was seen as embarrassing]. And that was so surprising, that genital warts was what they cared for, it did not match [with the general idea that cervical cancer is the most important part].

Emma emphasized that she believed that a focus on genital warts would increase the vaccination coverage. By emphasizing that the age group is relatively sexually active, and that the girls themselves cared for genital warts as they were embarrassing, her statement complicates the idea that girls are at an age where it is inappropriate to address them through a focus on sex. Emma continued by adding "it is not only [Bredland County Council] that does [exclude sex] ... Also pharmaceutical companies have launched it as a cancer vaccine". Emma continued: "And there's so much shame around infections compared to cancer. Cancer, you can be a martyr and 'poor you' and 'keep strong!' But infections are about risk affecting other people and ... It's really interesting how one does infections versus cancer". Stating that cancer is valued higher than infections, Emma claimed that the decision to exclude sex had to do with a general idea of cancer as a proper disease, and (sexual) infections as shameful since they are seen as having to do with how individuals' "risky" behaviors may affect others.

Emma stressed that even if the girls wanted to get vaccinated to avoid shame, to include genital warts was not assumed as a problematic burden by the girls in the interview study she had conducted. If the idea, as discussed by Klara, of lifting the burden from girls was about caring for the girls, Emma's answer invokes this care as problematic in a vein similar to how the county council's interview evaluation was discussed in the interview with Klara. For the girls, Emma meant, it was good that they could get vaccinated to avoid such shame. Thus, even if Emma positioned herself as critical toward a discourse enacting sexual infections such as HPV as shameful, in her words, Gardasil was invoked as a matter of care for the girls as it could help them avoid shame and embarrassment, which was something *they* cared about. This is reminiscent of Klara's emphasis on new data that made it evident that they needed to change the campaign focus as girls *themselves* problematized the idea that including sexual content would be problematic.

In both Klara's and Emma's answers interview studies with girls are invoked as methodological devices that helped them argue for the need to *listen to girls* as an important matter of care. Indeed, Emma's focus on how her study showed that girls cared about genital warts is similar to Klara's use of the interview evaluation to emphasize that making the sexual dimension, disease and death absent did not work. They both made use of these studies to articulate what girls *themselves* care for.

How the two interview studies discussed by Klara and Emma complicated and problematized how the county council cared for girls can be understood as what I introduced in Chapter 6 as care troublers. Additionally, in staging failing or insufficient matters of care by providing alternative ideas for "good care", they were simultaneously invoked as care enablers. That is, the interviews as methodological devices troubled one version of care by providing an alternative that was articulated as more in line with what the girls wanted. As devices that simultaneously troubled and enabled ways of caring for girls, the interviews' agential capacities to make present girls' own ideas were staged by my interviewees.

In doing so, the interviews brought up by Klara and Emma figured as a means of making girls active participants of care in a re-presentational manner. This is a specific form of public participation in technoscientific governance; the very aim of the two interviews was to increase vaccination coverage for this group.

Helena stressed that she did not agree about the decision not to include sexual behavior, as one needs to be able to talk about how HPV is transmitted. She explained that, when she as a communicator had talked with girls, they had often asked, "how does one get HPV?" Therefore, she asserted, one needs to be able to answer "from sex, through intercourse". That the county council, based on the interview evaluation, decided to change their campaign was therefore according to her, something good.

> And then we thought that genital warts actually is a message that suits younger people, this as cervical cancer is something that most often affects you further on in the future. If you're 15–20, then perhaps you don't think "shit, I can get cervical cancer when I'm 40" [...] Genital warts are closer in time. Perhaps you as a teenager don't protect yourself [when engaging in sexual activities].

In this excerpt, age is stated as a reason for including genital warts. In striking contrast to Klara who earlier stated that a reason for not including sexual behavior was the girls' age, age is here emphasized as the very reason for doing so. Genital warts, it is stated, affect girls *now*. Instead of being located in a distant future, genital warts are presented as something that teenagers care for in the present. This is reminiscent of Emma's discussion around genital warts as good to include as this is something girls care for *now*.

Making present genital warts as affecting girls now is interesting in how it does *not* present HPV vaccination first and foremost as an anticipatory immediacy. When Emma and Helena did trouble the absence of genital warts in the first "I love me" campaign, they also, implicitly, problematized anticipatory immediacy as a suitable mode of address. Instead of a message about the future, they felt that a message about teenagers' contemporary life would be fitting. For this catch-up group, according to Helena and Emma, sexuality is already present, and therefore it makes sense to address them with reference to that. Even if a mode of address focusing on genital warts as a matter of care about the present by both Emma and Helena implicitly had the main goal of enabling an increased vaccination coverage to prevent future cancer (as anticipation), their answers stressed *present genital warts* rather than *future cervical cancer*. This opened up a possibility of articulating HPV vaccination as something other than as anticipatory promises of future health, and unsettled a given status of HPV vaccination as being about anticipating cervical cancer.

Facts, feelings and Facebook

Klara emphasized several times that they did "include all the facts" in the campaign. For example, she stated that even if genital warts were not included in some parts of the campaign, it was part of the fact-based pamphlets sent to parents. She continued:

> So it's really a combination of facts that make it possible for parents to make a decision, this especially if the target group has not come of age. Thus, as part of the design was both strictly fact-based parts, where we don't work with … what should I say? … intention impelling messages, almost at all. [There are also] parts [where] we do work with intention impelling messages, for example "take care of yourself this summer, get vaccinated". That was explicitly something intention impelling. But it's important for me to highlight the combination. A [communication] strategy is a combination of messages. […] The different media are integrated.

It was by Klara stressed that all the facts were indeed included as "[t]he different media are integrated". Therefore, the county council did not elect for either facts *or* "intention impelling messages" about care and love. At the same time, a separation between "intention impelling messages" and "fact-based messages" was articulated. Later in the interview Klara stressed this message about care as a "message about love" and, in general, she talked in the interview about differences between fact-based campaigns and feelings-oriented ones. An idea of care and love as something that can be used to impel girls to get vaccinated was invoked. In separating this from, as it was stated in the excerpt above, "strictly fact-based parts", a separation between factual messages and affective messages about care and love was performed.

In addressing parents through "strictly fact-based" messages, and the girls through affective messages of "I love me" or "take care of yourself this summer!", age and gender matter; facts suit parents, feelings suit girls. This differs from how I earlier in this chapter discussed how Emma and Helena stated that getting vaccinated should be the girls' own decisions. Different articulations of how girls make decisions circulated: girls as knowledgeably and rationally choosing subjects, or girls as affectively driven. Moreover, Klara's reasoning includes a view of parents: that parents will act based on facts and not feelings. This view differs from how sociological vaccination research has discussed discourses articulating parents as using affective and experience-based knowledge as the basis for making vaccination decisions (Hobson-West 2003; Leach and Fairhead 2007).

Girls as linked to feelings were not the only articulation. During other parts of the interviews, girls as the ones who ought to know the facts was explicitly emphasized. For example, as an answer to a question about how they respond to vaccination critique, Helena said that to girls they "explain that everything we do is based on scientific facts and that it's up to each individual to learn as many facts as possible and to make an own decision based on what one thinks is the best for oneself". In response to a question about what information should be communicated, she also indicated girls as the ones who should learn the facts. These "facts" were:

> [That] the vaccine protects against 70 percent of the HPV viruses that exist. That we're also always explaining that it does not give a 100 percent protection and in almost all places we also say that it's important that one keeps on taking the Pap smear from 23 years old. As if you do that, both get vaccinated and take the Pap smear, then you have the best possible protection against cervical cancer available one can provide for oneself. Thus, it's really important to explain that one is not completely safe after having taken the vaccine.

In this excerpt, Helena stresses some of the often emphasized specificities regarding HPV vaccines. In doing so, she emphasizes that it is important to provide information that makes it clear to girls that they will not be fully protected. At the same time, not just any evidence was drawn upon as "good information". Boundaries were drawn between medical facts and other claims. Helena continued:

> We would not, you know, refer to what Mothers Against Gardasil say [laughter]. Not we at least. But then all the others can do it ... On the Facebook site, for example. And that they do sometimes. But we don't. Instead, for example a study conducted by the *Karolinska Institute* last year has been really good to use.

The Facebook site is in this excerpt brought up as a way of delineating facts from opinions, something that was common in the interviews. Helena continued by stating that on Facebook "whoever can say whatever". "Everyone is allowed to think differently but sometimes it gets a bit ... Perhaps it isn't really things that are in accordance with our main aim that are posted on Facebook", she continued. In a related vein, Klara stated that communication on the Facebook site was not in line with the "vaccination goal":

> [D]uring periods, it was many girls that were active and active in a way we wanted […] [T]hey could communicate with each other on this theme and they could ask us questions. "Where do I get vaccinated?"; "Is it dangerous?" That is, dialogue-based parts of the communication that is supposed to work as a support for the girls. We could see that during some periods it was like that, but during long periods it was mostly disorder and a mess of different groups of people with different opinions that talked with each other. And that did not support the vaccination goal at all […] It wasn't at use for the target group […] It became an arena for mud throwing.

Emma talked about the Facebook site in a manner similar to Klara and Helena, and she stressed that the opinions raised on Facebook were problematic. "I can just think that it becomes so biased that it's not balanced and, you know, scientific […] It's a shame that it's possible to distort science like that," she said. These statements from Klara, Emma and Helena are significant. Disorder and mess, mud throwing and Mothers Against Gardasil become invoked as matters of *concern* as these things complicate the possibility of having a dialogue with girls about matters that are in line with the aim and goal of the county council, and that is helpful for the girls. Whereas the county council wanted a dialogue where people could agree upon the matters to be discussed, people on the Facebook site did by no means agree upon what these matters should be. Klara's, Emma's and Helena's statements invoke the idea that instead of participating in the dialogue hoped for by the county council, people used Facebook devices to spark a vaccination controversy. Importantly, this was articulated as making it impossible for the county council to care for the girls on the Facebook site.

Care as being where the girls are, and as learning from girls

In the interviews I conducted, the pink vaccination trailer tour and the trailer tour were discussed (see Figure 6, page 128). For example, Emma stressed that the trailer was "really effective, popular, really, really good" as "it really has that common touch". With "common touch" she referred to an idea of vaccinations as being about people waiting in a long line for the shot, and to the blood buses that in Sweden are used to encourage people to donate blood. She continued by discussing the trailer in terms of care accessibility:

> And this with [care] accessibility, too. It's the easiest … If we would have [a trailer] arriving just outside your door [or school] and that would offer something for free … Of course you would then wait outside and perhaps decide to get vaccinated, something that perhaps you wouldn't do otherwise.

In Emma's words the vaccination trailer was a success as it provided care accessibility. In a similar vein, Klara related to care accessibility when I asked why it was decided to include a trailer in the campaign. She answered:

> It's really about creating [care] accessibility. This as the target group, and this goes for both vaccinations and other things, has a lot on its plate. And when you're young perhaps you don't understand that … the planning horizon that a scheduled vaccination appointment at your house doctor in two weeks … I can have forgotten about that one. But if I know that I have [care] accessibility close to me. "Yes, this about the vaccination, I can go around the corner". Then it becomes a different thing. That makes it is easier to take that step. So we just wanted to increase the accessibility to communication, to vaccination.

Klara and Emma stressed the trailer as a way to make it easier for girls to go and get vaccinated. Girls can get vaccinated just "around the corner" or just outside their door. This was also stressed by Helena who said that the trailer tour "was a concrete intervention that we did to [increase care accessibility]". She continued: "We went out and were there. And we made it easier for them [the girls]. We really did what we could to make it simple."

Emma, Helena and Klara stressed the trailer as a way of increasing vaccination coverage since it brought care closer to where girls are. As such the trailer figured in the interviews as a care enabler that brings care to girls. This stages the trailer as reminiscent of a regular vaccination setting where the vaccination "comes to" the children at school. Care becomes *mobile care*. In the interviews the trailer was not only discussed as materializing accessible care, but also the message of "I love me". This is reminiscent of how Lotten Gustafsson Reinius with colleagues (2013) discuss different propaganda and health care buses through a reworking of media historian McLuhan's (2009 [1964]) famous statement that "the medium is the message" into "the bus is the message". In the interviews, the trailer was the message.

With the trailer the county council wanted to be where the girls are, and thus enable care accessibility. The trailer as an example of how the importance of being where the girls are was also brought up in the interview with Klara. She said: "If the county council should be able to reach out with a message, then it is our obligation to talk with all the citizens. And then we need to reach them where they are." Reaching girls where they are was thus staged as an important matter of care; it enabled accessible care.

For Helena, being where the girls are was also about listening to and learning from the girls. Being where the girls are means that they can respond to what the girls express they need. That is, it was not only about accessible vaccination. She continued:

> We learned a lot from talking with the girls and hearing their thoughts. We discovered that many girls are really scared of injections, it was so many complaints and exclaims in the trailer sometimes so you wouldn't believe people that it was true! So that is really a typical example of care accessibility then.

Helena emphasized how girls reacted to the trailer tour. She especially focused on how many girls were anxious or scared of injections. This was seen as a question of care accessibility. Due to the county council staff *being there* they learned that girls are scared of injections. In this way, in the words of Helena, care accessibility, as *being there with the girls*, was an act of care for girls. Girls' well-being was the matter of care, rather than the county council's need to increase the vaccination coverage. Care became a matter of responding to, and attending to, situations of worries, anxieties and fear. Several times during the interviews, my interviewees stressed that during situations when girls were afraid of injections it was important to take them seriously and listen to them.

This focus on listening to the girls is partly reminiscent of how Emma and Klara stressed how the county council's interview evaluation, and the interview study conducted by Emma, illuminated what girls *themselves* cared about. In these examples, a predominant focus was on learning from girls what they need and want. That, indeed, was articulated as a vital matter of care. However, while the interview evaluation and the interview study were explicitly discussed as ways of listening to the girls by means of learning from them how it is possible to get more girls vaccinated, the situation of listening to girls in the trailer, and responding to their anxieties and fear about injections, was not simply discussed as ways of learning from girls how to get more girls vaccinated. Instead, the fact that they were scared and anxious was staged as central. It was stressed that it was important to make girls understand that it is not something dangerous to get vaccinated but, at the same time, that it is alright to feel scared or anxious. That is, listening to girls was not only a mode of governance to improve vaccination; it was also about responding to, and taking seriously, unhappy feelings.

It was emphasized that the trailer transformed vaccination from a difficult matter where girls need to find, and go to a vaccinator, to accessible and instant care. Locating care to where girls are was envisioned to turn it into an easy and quick task. That is, the trailer was envisioned to *speed up* care, and to enable an intensified temporality of care. In this way, the trailer was articulated as a *pacing care device*. This differs from how Puig de la Bellacasa (2015) discusses "the pace of care" as an alternative "care time" that may challenge futurity. The trailer as speeding up care is about a vision of immediacy that may tap into a trope that getting vaccinated is something to do *now* to prevent future disease. That is, a predominant vision of the trailer was that it *enabled* anticipatory immediacy as it materialized the message "get vaccinated now".

Yet the trailer also came with a pace of care other than an intensified speed of vaccination. Instead of simply enabling intensified care, the trailer also served as a *location of* care that inhabited, and enabled, an attention to anxiety and fear of injections. Such situations of fear and anxiety likely slowed down the pace of vaccination. Since it was emphasized that it was crucial to listen to, and take seriously, girls' fears and anxieties, an intensified care was not the only temporality present. A slowed down pace makes space-time differently, illustrated by how the trailer made possible "an active listening, an opening up for surprises" (Schrader 2015: 673). This articulates an alternative pace of care which, through a link with fear and anxiety, slowed down the speed of vaccination, and which included a care for fear and anxiety rather than the "I love me" feelings of love and happiness. Paradoxically, this "care time" was provoked by the very setting of the trailer tour as a practice for speeding it up.

Conclusions: caring for, caring with?

I have discussed some paradoxical links between girls and care. I have focused on how the "I love me" message was discussed by my interviewees as a matter of encouraging and empowering girls to care and love themselves. It was about empowering girls to empower themselves. At the same time, a "care for the herd" was invoked as a way of troubling this message. However, a focus on the individual girl was "winning" over a care for the herd, as the herd was being linked to a question of compulsion. When it became a question of choice versus compulsion, choice was emphasized as the right mode of address.

An important matter of care had to do with listening and responding to what the girls *themselves* cared about. This was done through different devices: an interview evaluation, an interview study and the vaccination trailer were enrolled to enable this matter of care. In addition, when people other than girls occupied Facebook, it was invoked as a matter of concern as it did make it harder, or even impossible, for the county council to listen to the girls via this arena.

I have discussed the methodological devices of the interview evaluation and Emma's interview study as care troublers that staged the ways the county council cared for girls as failing or insufficient. As feminist STS scholar Ana Viseu (2015) shows, attending to failures in how care is configured can highlight how care not always is something desirable. As she also shows, focusing on failures also enables attention to how care can be done differently. By attending to how the methodological devices served as simultaneously care troublers and care enablers, I aimed to point at this coexistence of trouble and generative action: trouble *generated* transformation. More concretely, in working with the idea of devices as care troublers, I have discussed how the decision to exclude a sexual dimension from the campaign was problematized by the majority of my interviewees on the basis of their interview evaluation. Their evaluation articulated that their way of caring for girls was not in accordance with what the girls actually cared for. As a care troubler, their interview evaluation did articulate their former care for girls as a *failed* matter of care. The device was enrolled in the interview to problematize the idea of care as protection. Protection, my interviewees articulated, turned out to be a matter of speaking for the girls without listening enough to what they wanted. In a similar vein, Emma drew upon the interview study conducted by her to problematize the absence of genital warts by stressing that what girls care for is actually not cervical cancer as something that can happen in the future, but, instead, genital warts as an issue affecting them in the present. Both Emma and Helena articulated a version of HPV vaccination other than one that has primarily to do with anticipatory immediacy. Instead of as a promise for healthy futures, it became a matter of avoiding shame and embarrassment affecting girls now. This differs from predominant articulations of HPV vaccination as an anticipatory practice.

Seeing the methodological devices as care troublers deepens the discussion from Chapter 6 on care devices. In a similar vein to the Facebook devices that troubled the "I love me" vision of care as love and happiness, these methodological devices were in the interviews invoked as data that illuminated the problems with how the "I love me" message was configured. However, whereas the Facebook devices participated as part of collectives along with diverse human subjects, the methodological devices were used to improve girl-centered public participation. In fact, in the interviews the Facebook devices were articulated as matters of concern as they enacted care that was *not* girl-centered. That they did not facilitate girl-centered care was invoked as the very reason for why Facebook did not work as an arena for public participation.

Listening to girls through methodological devices such as a county council interview evaluation is a specific, choreographed form of listening. Such devices serve as "formalized mechanisms of voicing" (Michael 2012: 531). In the context of the first "I love me" campaign, as devices for re-presenting girls' opinions, they elicited what was perceived as relevant opinions for increasing vaccination coverage. That is, even if the county council changed how they cared thanks to the devices, they still worked with devices that inhabited the idea that learning what girls "out there" really think will increase vaccination coverage and that this can be done through highly structured evaluations that enact clearly defined answers, and clearly defined girls. As such they inhabit a very specific mode of attention. In my interviews with the professionals at Bredland County Council, they figured as attempts to care *with* girls, which provoked specificities and trouble in how the county council designed girl-centered public participation. This trouble was mediated through these methodological devices, designed for eliciting girls' opinions. As such, they were what STS scholars Javier Lezaun and Linda Soneryd entitle "technologies of elicitation". Such technologies "seek to engage publics in dialogue and generate certified 'public opinion' with the ultimate goal of increasing the *productivity* of government" (Lezaun and Soneryd 2007: 282, emphasis in original). Therefore, they provoke "a highly formalized and carefully choreographed form of engagement" with a purpose "to generate a stable referent" such as "public opinion" (ibid.: 282). As technologies (or devices) of elicitation, the interview evaluation and Emma's interview study (as methodological devices) figure as modes of governance that choreographed *how to* listen to girls since they entailed a clear goal: to increase the vaccination coverage.

The methodological devices for listening to the girls used by the county council can be easily critiqued for entailing a mode of listening that Schrader (2015) discusses as a progressive temporality of learning (learning as an enlightening and goal-oriented activity). Through inhabiting a clear end (increasing the vaccination coverage) and already-defined subjects (empowerment girls) care becomes a speaking for the other as (direct) helping action ("get vaccinated now!"). Such a mode of listening easily closes down possibilities for indeterminacy and surprises as it provokes clear-cut answers re-presenting "public opinion". However, as Lezaun and Soneryd (2007: 295) also emphasize, technologies (or devices) of elicitation can be understood as "engines of movement" that often have unanticipated events or dimensions. Paying attention to how the interview evaluation and Emma's interview study, even if their ultimate goal was to increase vaccination coverage, also served to problematize a predominant assumption about girls as in need of protection from "sex negativity", it is possible to see how they did more than simply aim to improve the vaccination coverage for this group. As care troublers they did more than simply reaffirm a need for vaccination. Instead, a matter of care other than the predominant one was made present, something that served to illuminate (some of) the complex politics in how girls can be cared for, and in how they care.

The same can be argued about the trailer. As a device for girl participation, it became a question not only of fully choreographed strategies. It was not simply about enabling, speeding up, and reaffirming, vaccination, but also about responding to fears and anxieties. As such the trailer figured as an engine of movement (in a literal sense!) with unanticipated dimensions. As Schrader (2015) shows, such movements do not have to mean progressive and intensified time, but may include a slowing down, and an opening up for surprises. The trailer tour staged a situation that included a coexistence of different temporalities of care, something which shows how health communication does not have to be a matter only of immediacy (such as direct helping action) and already-defined action.

The trailer tour illuminates insights about the spatial dimensions of care. The county council aimed to intensify or speed up care through the trailer. Still, as a spatial location that inhabited a setting where girls' fear and anxieties of injections held center stage, the trailer, and the trailer tour, staged another temporality of care. The trailer tour as a location for care inhabited two coexisting temporalities of care: anticipatory immediacy as inhabiting a vision of intensified or instant care and fears and anxiety as slowing down care. This means that the trailer did not include a predetermined agency. When it got linked to girls, it was articulated as a location for taking seriously girls' fears and anxieties. Being where the girls were, and listening to them, turned out to include more than enabling intensified care for the future. It enabled a slowing down of care. Importantly, caring *with* girls was also about being there with them and responding to the needs of that very situation.

EMPIRICAL PART III CARING ABOUT CANCER STORYTELLING

Cancer stories are often affectively dense. Some invoke promises of success, happiness and survival, others tell narratives about grief for a diseased loved one, or about fear of cancer tumors having spread in the body. Stories from patients or their relatives often invite the audience to care for their story and, in addition, to care for cancer as a circulating trope and reality. Empirical Part III of this thesis zooms in on this phenomenon by focusing on a particular form of cancer stories: cancer narratives figuring in health campaigns for HPV vaccination. More concretely, it discusses the cancer narratives in the second "I love me" campaign. It consists of two different chapters: "Affective Relationalities in Cancer Storytelling" (Chapter 8) and "Communicating Death and Disease with Care" (Chapter 9).

The second "I love me" campaign retells cancer narratives from cervical cancer patients, and from relatives. In the campaign information from Bredland County Council, it is stated that they "make visible authentic testimonies" (campaign information from 2013, my translation from Swedish). This mode of address was decided on based on an evaluation the county council did of the first "I love me" campaign (discussed in Chapter 7). Based on this evaluation, the county council decided that the campaign "message was in need of getting more dramatic to really make people understand the consequences of this life-threatening disease"

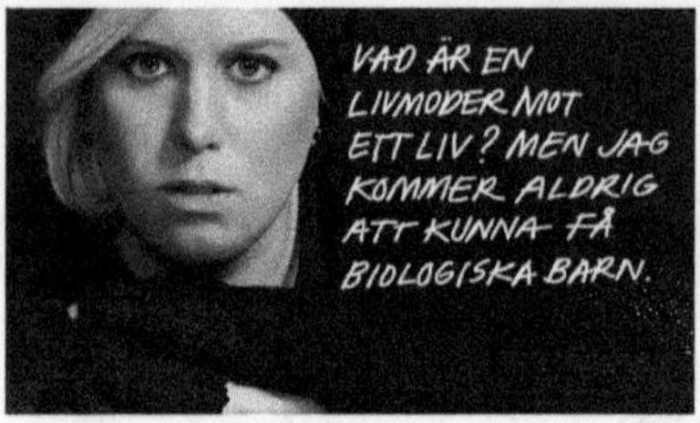

Figure 7. "What is a uterus compared to a life? But I will never have biological children."

Figure 8. "The only thing I wanted was that I could be sick instead of her."

(ibid.). This was because the evaluation showed that even if many people did know about HPV vaccination, the county council's vaccination goal of 95 percent of girls and young women vaccinated was far from reached. For the county council, it was clear that a more dramatic campaign strategy was needed.

The second campaign was launched after the decision was taken by Bredland County Council to extend the target group for the catch-up vaccination from girls from 13 to 20 years old to girls and young women up to 26 years old. Therefore, as against the first "I love me" campaign, the second one was directed toward girls and young women from 13 to 26 years old.

As already mentioned in the methodology chapter (Chapter 3), the campaign narratives were published online on a campaign site. Moreover, during two different campaign periods (May and September 2013) campaign images depicting excerpts from the cancer narratives, together with images portraying the storytellers, were re-presented in posters within the public transport system (for examples, see Figures 7 and 8). As with the first "I love me" campaign, the stories were also posted on the "I love me" Facebook site. How the stories are articulated on Facebook is not, however, focused on in Empirical Part III.

The narratives in the campaign re-present cancer experiences in specific, partial ways. The county council found the storytellers on blogs where they had been blogging about their cancer experiences. These bloggers in turn agreed to be interviewed by the county council. The interviews were transcribed and edited at a later stage by the county council to become campaign narratives. Thus, the cancer narratives told by the storytellers were *translated* into campaign cancer narratives.

This means that they include specificities related to a chain of translations *ending up* with cancer narratives in a campaign. At the same time, and as will be discussed in Chapter 9, to ensure that the narratives were "authentic testimonies", the county council emphasized that the wording of the storytellers was edited as little as possible.

In the first chapter of Empirical Part III (Chapter 8), I focus on how the campaign mediates cancer experiences, and takes part in invoking different affective responses. Importantly, I discuss how a *retelling* of cancer narratives can be done responsibly by attending to cancer storytelling in the format of campaign narratives as including multilayered matters of care. In the second chapter (Chapter 9), I attend to how my interviewees discussed the campaign. In particular, attention is on how boundaries were drawn in the interviews for when, and how, a public organization, like the county council, can responsibly communicate a public message about death and disease. Thus, I engage in a discussion concerning links between public accountability, responsibility and care. The chapters in Empirical Part III concern different dimensions of, and links between, matters of care, temporalities, responsibility and accountability within a context of cancer storytelling about death and disease.

8. Affective Relationalities in Cancer Storytelling

"[T]hey really just use a tactic of scaring people", my friend wrote to me in the fall of 2013. Attached to her message was an image she had taken on the public transports system that depicted a black-and-white image from the second "I love me" campaign. It was a photograph of a woman looking serious, devoted and determined. The woman's gaze was intense; she stared into the camera and out to the viewers. Underneath the image was the text "What is a uterus compared to a life? But I will never have biological children" (Figure 7, page 178). Looking at the image, it felt like she looked at *me* and I could not help looking back. I replied to my friend and stated, "Yeah, I know! It's striking and somewhat problematic how they try to scare people into getting vaccinated".

By looking at the image of the woman and reading that she never can have biological children due to cervical cancer, it struck me that the point was that the viewer of the image *should* be affected by looking at it. The viewer is encouraged to feel that, "that could become me in the future" – and thereby decide to get vaccinated. Perhaps not surprisingly, health campaigns similar to this one are often criticized for being about scaremongering, in that they try to scare people into changing their behavior to become healthier (Gagnon et al. 2010; Lupton 2013; Porroche-Escudero 2014). One could argue that the second "I love me" campaign is yet another example of scaremongering, an obvious case; nothing more to say. Yet I will explain that it is not that simple.

After my friend's short text message, I felt uncomfortable about my reaction, about my response. It felt like there was more trouble than fear going on here, and that it could be important to stay with my discomfort as something potentially generative that could open up another engagement. Was the sense of unease that the campaign awoke for me and my friend – and our sense that a critique was needed – based only on our concern about the campaign invoking fear? Is the campaign – as I stated in the text message – "striking and somewhat problematic" merely because of that? Because of my discomfort about our conversation back then in 2013, I went online to reread the narratives. In one of them Ida tells me about her story about her mother's cancer and her reaction toward it:

> It started with my mom feeling really hard in the stomach. I remember that I was in the bathroom and I heard mom and dad talking about it. Words like "gynecological emergency ward". And then everything happened really fast. My mom had cancer. I went to junior high school and that was already a difficult time in my life. Who am I? What shall I do? How is my life going to be? And then this happened [...] A strong memory I have is from when my mom went to the hospital. Dad and I ate dinner at home, in silence. When we were about to go to bed, we found a letter from mom on our pillows. Something like "take care of yourself, I hope you'll be okay, I think about you and I love you". I fell apart. It was like a farewell letter. I tried to manage school at the same time but a lot of people noticed that I didn't feel well. I turned in on myself and pushed the pain away, even for my friends and family. It felt better to take a step back and to leave the ones I loved, instead of having them taken away from me.

In this narrative, Ida's affective response toward the cancer is in focus, rather than it merely being on the cancer diagnosis; it is a story about memories of fear of death and of Ida's love toward her mother. Relationships between Ida and her diseased mother, Ida and her father, her parents and Ida and her mother's cancer are focused. Ida's narrative, and several of the others that the second "I love me campaign" consists of, affected me. I was moved by them. When reading Ida's narrative, I could not help but think and feel that her story was somehow also my own story. Therefore, I will retell a story of my own.

When I was 13 my father died of cancer, after six years of treatment. How Ida's reaction is described in her narrative resonates with how I reacted after my father's death – and during the many years of his sickness. As I remember it today, all I wanted then was to act like nothing had happened. I did not want anyone to talk to me about it. That made it way too real. I just wanted to be a normal teenager; to *pass* as a normal teenager. Not the one with a sick, and, later, dead father. As stated in Ida's narrative, I also "went to junior high school and that was already a difficult time in my life". These could have been my words. For me, the urge to appear normal, instead of being the girl whose father was sick, and then, later, dead, made me not talk at school about the fact that he was sick nor that he had passed away.

The day after his death I went on a camping trip with my new high school classmates, without telling anyone about the situation (however, eventually I told my closest friends during the trip – it just felt too absurd of me not to). It felt like I could not stand staying at home. Being at home with a family in grief made my father's death (and perhaps even more, my family's pain and grief) far too real, too close. I needed to keep it at a distance. I felt a strong desire to talk about regular teenage stuff. That would hold the world together, I felt. Being the girl with the dead father was too scary and shameful. In the following months, I hesitated to invite friends home, as I did not want people to talk to me about his death. Like in Ida's story: "I turned in on myself and pushed the pain away, even for my friends and family". Reading Ida's story invoked these highly affective memories; I felt the pain, the grief, the shame and the anxiety again. Thus, in line with the campaign's intention, in the meeting between me and the campaign, an affective response was generated. However, it was not merely fear that I felt; a range of different feelings such as grief, pain and shame were evoked – including bodily intensities. Yet the question remains, how can one analytically respond to a health campaign that can invoke such painful memories?

In this chapter, I try to respond by engaging with possibilities for responsible re-storytelling. I will use the notion of *affective relationalities* (Jerak-Zuiderent 2014) to account for the narratives told and the feelings – and worlds – they inhabit and generate (my own affective responses included). I attend to matters of care on two dimensions: matters of care re-presented in, and invoked by, the campaign narratives *and* re-telling these narratives as a matter of care in itself. In doing so, I will affectively engage with the campaign's relationality as a way of making others care (both as critique and affirmation) about its existence and for possible future becomings of health campaigns as caring.

I aim for "affective engagement rather than *directed* emotional responses" (Jain and Stacey 2015: 15, emphasis added) as this puts the attention on processes of becoming affected by, and affecting, the texts and worlds I engage with in this chapter. I focus on affective relationalities to illuminate that how cancer stories are retold matters for how responsible critiques can be performed. In wanting to retell the campaign stories responsibly, I aim to take seriously the mediated memories and tropes of feelings such as pain, loss, grief, love, and hope that circulate in them and the "affective cancer trouble" (Lindén and Sullivan 2015: 14) these re-present and provoke.

Trying to engage with the narratives responsibly means staying with the trouble of such engagement. Returning to my affective response when reading Ida's narrative, my response is not necessarily desirable and it is definitely not innocent. Importantly, Puig de la Bellacasa (2012: 209) highlights that all too easily caring "can lead to appropriating the recipients of 'our' care, instead of relating ourselves to them". This means that attending to care as *sameness* and *identification* (as I did: "it felt like Ida's story was somehow also my own story") risks letting my cancer story take over Ida's. Therefore, it becomes important when trying to retell Ida's narrative – and the other campaign storytellers' narratives – together with my memories responsibly and carefully, and in a way that does not obliterate Ida's narrative but does not fetishize it either. Moreover, these cancer narratives are not just any narratives; they are part of a vaccination campaign. As I wrote in the introduction to this part of the thesis, they include specificities in how they are the result of translations *ending up* with these campaign cancer narratives. Therefore, it becomes crucial to attend to the troubles and promises that the campaign includes and invokes. In doing so, I attend to how affective memories as matters of care are mediated textually and visually and, as part of that, how the narratives make present certain versions of, for example, fear, love and pain.

I start with a discussion of the campaign's visual re-presentation, and then move on to the textual. When analytically fruitful, I will weave together the narratives of the campaign with my own memories of my father's disease and death. In doing so, and as explained in Chapter 3, the chapter engages an "implicated reading" as a commitment and attunement to affective relationalities.

Images mediating cancer experiences

In contrast to the first "I love me" campaign with its colorful (often pink) message about healthy, happy girls, the storytellers in the second "I love me" campaign are imaged in black-and-white against a black background (Figures 7 and 8, page 178). This draws on certain tropes: black-and-whites are often used to denote something being more real and serious (Stein 1991; Hariman and Lucaites 2007) and signifying disease or death (Cooter and Stein 2007). Playing on such tropes, the black-and-whiteness of the images mediates the stories told as being about cervical cancer as a serious matter of death and disease.

In the images, only the faces and torsos of the storytellers are present, and, therefore, the storytellers' facial expressions are in focus. All storytellers look directly into the camera and, thus, out to the audience.

There are no smiles, just serious expressions. The storytellers' staring eyes are in focus; their intense gaze invites me to look back, to take part of their story. It is almost as if their grief, their sorrow and pain is imagined only through their eyes (something that invokes a body–mind dualism: instead of re-presenting disease and death in the body it is invoked as a matter of the mind; a trope is invoked of the eyes as the mirror of the soul). The visual re-presentation of their eyes presents them as in control of themselves and of their story, and a specific matter of care is made: viewers are invited to care for *their* stories.

Instead of the storytellers being portrayed together with the relatives they talk about, all the images, except one, portray them alone. They stand alone without *anything else* (subject or object). It is just them in black-and-white against a black background. Displaced from their everyday life and from possible relationships, the imaging of the storytellers invokes absence. In the textual narratives relatives are talked about with signifiers of specific relationships: "mom", "her", "my partner", "my sisters". But in the photos these relationships are visually, strikingly, absent.

Death and disease are also visually absent in the photos. Thus, even if it can be argued that the black-and-whiteness of the images in itself denotes death and disease, without the text related to each storyteller this interpretation would not be possible. The visual re-presentation of the bodies of the storytellers shows no sign of trauma. This is also true for the two ex-patients re-presented in the images. There is no sign of past disease. Instead, visually, the bodies re-presented look strikingly healthy.

The re-presentation of healthy bodies can be related to a trope of an ideal able body in control of the disease. In connection to cancer this trope invokes the cancer body as not being bodily affected by cancer. It is important not to equalize *seeing* a cancerous body with *knowing* the disease, as cancer tumors can grow inside the body without visible signs of disease. Still, such a healthy cancer body trope can, as for example Cartwright (1998) and Jain (2013) discuss, be productively contrasted with feminist and critical art work where chemotherapy-marked and scarred bodies have been re-presented to trouble how cancer experiences are often re-presented in the public arena. In the "I love me" campaign, in contrast, cancer bodies are re-presented as healthy and strong bodies, as unmarked bodies. In this sense, this invitation to care for the cancer stories comes with a non-innocent and conditioned version of care; one that presents able, strong and unmarked bodies as what one ought to care for.

However, as is common in health campaigns, text and image work together to signify that they, or one of their loved ones, is sick, was sick, or is dead. For example, the storyteller Cecilia is portrayed together with the following excerpt from her narrative: "What if it has spread through my body? What if nothing can be done?" Here, a subject with a body that it is not possible to control is articulated. It is a narrative about fear as a matter of anticipating future death. Without Cecilia's knowledge, cancer tumors may have spread throughout her body. This message stands in contrast to her portrait, where she looks steadily and intensely into the camera. The fear present in the wording is what makes it possible for me to interpret it as a cancer story and as a story about not being in control of one's body and disease; a story about things happening in the body that the mind does not know about. The textual narrative makes Cecilia's story similar to how feminist scholar Jackie Stacey (1997: 67) describes people's fear of cancer as often being related to "a fear of something secretly growing inside the body". As in this example, in the second "I love me" campaign, the text together with the serious, non-smiling faces in black-and-white images direct the interpretation of the visuals as being about death and disease.

The words together with the visual focus on their eyes put the storyteller in the center. For example, in Cecilia's story she is the one telling us – mediated through the image centering on her facial expression – about her fear of the tumor having spread. This further amplifies the use of the gaze in the images. In looking straight into the camera and out to the public, the storytellers – Cecilia and the others – intensively gaze at the audience. The storytellers are performed as serious and determined; they are subjects of the gaze, of their stories. The audience is invited to see things from the storytellers' point of view; "this is my story, my experiences".

Playing upon a trope of photographic truth (Sturken and Cartwright 2009) – that the images give us unmediated access to the storytellers' experiences – the storytellers' facial expressions are related to their words. By use of the photographs, vision is invoked as crucial for how this form of storytelling is mediated: as an audience I am invited to look back – and care for – their narratives by their intense gaze combined with their words.

We all have a relationship with cancer

Several of the campaign narratives are about relatives' experiences. In Erik's narrative about his partner Kristina's cancer diagnosis, it is written:

> It was first when they called from the hospital and told me that they had to remove Kristina's whole uterus that I fell apart. All the anxiety was suddenly confirmed, concrete. Kristina tried to hold me, calm me, but no one was allowed to touch me. I broke down.

In this excerpt, the focus is on cervical cancer as a disease that affects relatives. Erik's story is similar to Ida's and mine, in that it is about how the cancer diagnosis made him stop communicating his feelings. In this sense, the narratives – mine included – are about a desire *not to relate.* To not be able to relate is, in Erik's narrative, described as *falling apart* and *breaking down.* In her story about her mother's cancer diagnosis, Ida talks about this too. Similar to Erik's narrative, she wrote that "I fell apart". *Breaking down* and *falling apart* depends on the idea of a whole human subject that, due to *too much* feeling, falls apart, breaks into pieces; becomes fragmented. Broken down into pieces – fallen apart – the subject cannot function and feelings take over. This builds upon a divide between control (or rationality) and feelings.

In these narratives, feelings are invoked as a force that takes over and makes it impossible for the subject to keep control. Keeping control here means *not relating*; not wanting people to calm you and to push people and the pain away. In both Ida's and Erik's narratives (memories of) affective experiences that signify a desire to be in control (but failing to do so) circulate. From Erik's narrative: "When we went to Uppsala for the operation I was in the worst place". It continues: "You have to be strong for both of you. Try to help your afflicted partner keep the mood up. The burden inside gets so heavy that you don't know what to do". Thus it is a narrative about Erik's experience of Kristina's cancer diagnosis. It describes the cancer diagnosis as an important part of the relationship; as something that has the capacity to transform their relations to each other. Thus, it is mainly a narrative about Erik's relation to Kristina, and how Erik felt that he had to be strong for both of them during Kristina's sickness.

Erik's narrative plays on ideas about male subjects not showing feelings, about being the strong ones that hold the family together through stability and rationality (and, importantly, keeping feelings inside all the way to breaking point), and to care through a holding-it-together management of feelings. But it is also a narrative about failing to do so, failing to cope as a stable, rational, holding-it-together male subject. "You want to protect each other in a family, but in a situation like this one, you feel like you are failing". In this sense, it is narrative about a gendered fear of failing to care.

The urge to protect others, but failing to do so, is also present in Ida's narrative. She describes anger toward her mother.

> I remember that my mom looked so small in the bed at the hospital after the operation. I just wanted to lay down next to her. But I was still so angry. It was like I thought "but can you please stop pretending to be sick and shape up".

As this excerpt indicates, in Ida's narrative *not relating* includes anger, and shame for feeling that anger. This stands in contrast to how caring for the other in Erik's narrative is, in a gendered way, articulated as not being able to hold-it-together. While Erik's narrative is about his failure as a male subject, Ida's is about her shame of not being able to show affection toward her mother.

Throughout Erik's narrative, cervical cancer is described as something that affects both men and women. The cervical cancer experience is articulated as a relational phenomenon that male subjects are involved in too. It is articulated as a gendered matter of care for male subjects in the sense of them caring for their close ones. Importantly, cervical cancer as a relational matter of care differs from how it is often re-presented as a women's disease. Erik's is not the only narrative where this is evident. For example, Ulrik tells us:

> First my big sister found out that she had precancerous cervical lesions. And then Ylva [his other sister] found out that she had cervical cancer. When your close ones are affected, you realize how important it [cervical cancer] is. You want to protect your sisters, but you are powerless.

Cervical cancer affected Ulrik's sisters and made him realize how serious the disease is. In addition, the excerpt describes how Ulrik has felt powerless, as he wanted to protect his sisters. The narrative does more than tell the audience what to do. It defines male subjectivity and (lack of) agency as well as affective reactions. It does this while presenting a male subject relationally tied to his sisters. Moreover, cervical cancer is re-presented as what makes Ulrik feel powerless – and what makes the feelings take over. The disease is also articulated as a condition capable of making Ulrik realize how serious cervical cancer is. It is, thus, a narrative that invokes a gendered feeling of (fear of) lack of control, and which simultaneously defines proper male subjectivity and depicts cervical cancer as a disease that invokes strong affective reactions.

Cervical cancer is in Ulrik's and Erik's narratives re-presented as a matter of care that affects not only women, or one that only girls or women need to take preventive action to avoid. It is articulated as a disease that generates strong affective reactions that are tightly linked to a rational and protective male subjectivity. HPV vaccination figures in their narratives as something that can save relatives from getting *emotionally affected by* cervical cancer (you don't want to fall apart, they implicitly tell us). The narratives tell us that you do not want to feel like Ulrik or Erik did and that therefore you should make sure your female partner or sisters are getting vaccinated. It invokes HPV vaccination as a matter of urgency: go home, encourage, push and spur your female partner or sisters to get vaccinated! In this way, their narratives are temporal in how they stage HPV vaccination as an urgency that requires us to *act now*, to *care now*. It becomes a matter of care *as* anticipatory immediacy.

As a relative of a person with a cancer story, reading the narratives I do not get the urge to immediately tell my close ones to get vaccinated. I do not feel like I need to *act now*. But I do feel simultaneously moved and troubled. Being a relative to someone who has died of cancer, it is easy to get upset about the possibility of other people around you developing cancer. That does influence – at least for me – how you relate to others' stories about cancer. Saying that I feel troubled and moved can be important; I think it is important to take seriously what cancer stories can tell us, even if they may upset us, even scare us.

In Ulrik's and Erik's narratives the emphasis is on their own experiences of their relative's cancer diagnosis. This is visible also in Lukas's narrative, the third of the three young men included in the campaign (Figure 8, page 178). My response to his story gives further insights into how getting affected by the narratives can be temporally more complex than what the idea of a matter of urgency envisions. His mother was afflicted by cervical cancer. From his narrative:

> My mom had cancer. It came as a shock, and my whole world stopped spinning. The only thing I wanted was that I could be sick instead of her. Mom had been through so many things already – it didn't feel fair. It became an extremely difficult time. To see your mom feeling that bad and being confined to bed for such a long time, that's hard. And then we didn't know how it would end.

In this excerpt, and similar to the other narratives, Lukas's affective reaction toward the cervical cancer diagnosis is highlighted. Lukas tells the viewer that his "world stopped spinning"; the cancer diagnosis is envisioned as disrupting and prohibiting the regular flow of life. This is a commonly used trope in cancer narratives (see e.g. Bell 2012). Moreover, he tells us that the cancer did not feel fair and that his mother was confined to bed for a long time. This (as do the other narratives) invokes an always, already presence of cancer; it happened to Lukas and his mother, it can happen to you or your close ones. Being described as *unfair*, cervical cancer is invoked as outside the control of the individual and therefore, implicitly, put in contrast to HPV vaccination as something that makes it controllable.

As a reader of Lukas's story, my own cancer memories were evoked. I started to think about how I feared the feeling that the world would stop spinning if I stayed at home, as I would then be confronted with the grieving family (I feared the moment when they would *break down*) and the realness of the situation. Reading about Lukas's memories of his mother being confined to bed made me think about how my father, during his last years, was so often confined to an armchair, diseased, and with blankets around his then skinny, pale body due to the cancer tumor's uncontrollable growth. I thought about the last time I met him at the hospital, two days before he died. My siblings and I were there to say goodbye and it was the first time I think I really understood that he was going to die. Reading Lukas's story invoked these affective memories; it invoked sadness, grief and pain.

What is more, it made me care for Lukas's story. It moved me to get affected by "the pain of the other" (Juanita Brown 2014: 181). In this way, the campaign not only turned cervical cancer storytelling into a matter of care for men, it also turned it into a matter of care for me as a viewer of the campaign. Through its affectively charged message it *related me* to Lukas's story and made me care about it.

My affective responses to the campaign open up questions about the ethico-politics of care. If Lukas invites me to hear his story, how to respond carefully – how to retell and look back – without projecting my story on his becomes important. How to get affected by "the pain of the other" (Juanita Brown 2014: 181) in a responsible manner in this context? In her article on care about deformed leaf bugs, Schrader problematizes caring that is based on "emphatic identification", as it requires a recognizable, and therefore already-defined, subject that one sees in oneself.

In other words, it requires self-recognition or self-presence, something that does "not put the subjects at risk" (Schrader 2015: 679). Emphatic identification, Schrader argues, relies on a vision of an intelligible subject that I, when I care for it, also categorize, stabilize and reaffirm. In other words, and as I mentioned already in the introduction to this chapter, it is a form of care that is based on an idea of sameness. Connected to this, to hear, as Ahmed (2004: 35) powerfully writes, "the other's pain as my pain, and to empathise with the other [...] involves violence". In other words, empathy based on sameness risks obligating and appropriating the other. Moreover, and as Schrader (2015) stresses, this form of care easily privileges humans or other animals that are culturally rendered easy to identify with.

Relating back to my care for Lukas's story, the affective memories and intensities generated when I read Lukas's story (how it made me care about his story) are partly about a feeling of identification as it is at least partly about me seeing myself in Lukas. Schrader's analysis illuminates how this is troubled territory. I do not know how Lukas felt; I only have his mediated story. I cannot put myself in Lukas's shoes.

The ease with which I relate to Lukas's story is not surprising. As Schrader (2015) indicates, playing on care as emphatic identification is a culturally powerful, and widely circulating, practice.[1] How, then, to respond responsibly when the situation includes potentially highly unsettling dimensions of power? Importantly, as a first, I care for Lukas's story in a different way than the county council with the campaign perhaps had planned. His story makes me care, but not automatically to act as a matter of urgency. How I relate to his story is not simply about a directed emotional response, meaning that it does not simply direct me in an anticipated direction ("get vaccinated now!"). It rather invokes a need to respond responsibly to his story. For me, it invokes a need for affective engagement rather than *directed* responses. This exemplifies how there is no straightforward relationship between the feelings invoked by the campaign and a decision to get vaccinated, and, as part of that, how

1. In her book *Moral Spectatorship*, Cartwright (2008: 2) proposes the notion of *emphatic identification* as an alternative to models of identification that are based on "the idea of feeling what the other feels" as about "imagining oneself as the other". She argues that emphatic identification is an often overlooked aspect of identification, and that it facilitates otherness rather than confirms sameness. This use of the notion is very different from Schrader's (2015); Schrader assumes that identification based on empathy is always about desiring sameness and to see oneself as the other.

getting affected is more complex and ambiguous than fully directed and anticipated responses.

One additional way of responding to Lukas's narrative – and to the other narratives – is to take seriously how it can tell us something about cancer and care (as part of health campaigns) rather than about Lukas himself. Responding to his narrative as an engagement with cancer and care directs the attention away from me imagining myself being Lukas, *and* allows space-time for the feelings circulating in the narrative. And importantly, as Jain (2013: 235n) argues, attending to such feelings "remain[s] central to any possibility of understanding the cultural status of cancer".

Working with my response to Lukas's narrative as an illuminating moment saying something about how cancer is cared for in contemporary society, it is possible to see how easy it is to relate to the campaign, as most people will have something to relate *with* (be it their own, or their close ones', histories of cancer, or stories they have heard – the circulation of cancer stories is vast). The campaign thus speaks to the idea that we all have a relationship with cancer. It is not hard to imagine that you or someone in your surroundings will develop cancer. In this way, the campaign speaks, simultaneously, to tropes and realities of cancer; people are always, already a part of cancer stories. The campaign plays on a mode of address where it potentially speaks not only to girls or young women but to everyone. Many people have, or can at least imaging having, a close one with a cancer history. Not everyone had a father who died from cancer, and in that sense, my story relates me to the campaign in a particular way; yet so do all stories, just in different ways.

Cancer reality as something always, already existing is articulated in the campaign. Focusing on people's relationships with cancer expands the possible number of people affected: me, Lukas, Ulrik, Ida. Suddenly we are brought into stories about cervical cancer. The campaign tells us that young women and girls have a relationship with cancer – but so does everyone else. It tells us that cancer *is here* and takes part in assembling our relations to our relatives, to cancer, to the present time and the future. Cancer as an affective state is about possibility and potentiality, it is about invoking the *it may happen,* or *it won't*. Its potentiality affects us in the present time. It is about the past telling us (telling me) that it can happen again, in the future. Cancer becomes a collective phenomenon that "becomes us" (Jain 2013) exactly due to the vast circulation of stories. Cancer is not merely mine, Ulrik's or Lukas's story. It becomes a highly collective, relational matter of care, which affectively aligns subjects to each other.

The care re-presented in the narratives is specific. Care is connected to temporality in how people are aligned to each other through articulations of cancer's potentiality. Moreover, and as I have touched upon already and discussed in Chapters 6 and 7, it is a vision of temporality that asks us to care immediately in an anticipatory and urgent manner: "get vaccination now!" As I have also already mentioned in Chapter 7, Puig de la Bellacasa (2015: 707) positions herself critically toward this link between care and time as an anticipatory immediacy. She writes: "Even when care is compelled by urgency, there is a needed distance from feelings of emergency, fear and future projections in order to focus on caring well". This critique against anticipatory forms of links between care and time is an important response to calls for urgency of action.

At the same time, learning from Jerak-Zuiderent (2014), it can be generative to slow this plot down as this allows me to take the "I love me" cancer narratives seriously in all their complexity. Rather than being simply about anticipatory immediacy, the narratives also inhabit and evoke a "care time" that is simultaneously personal, collective and temporally dense and messy. Evoking relational memories rather than only future-oriented anticipation the narratives *fold* different times and different affective relationalities. Their potentiality is also about the past. For example, and as my affective responses to the campaign illustrate, the feelings and affective memories circulating in and through the campaign are potentially multiple. Holding on to this, it is possible to stress how the narratives make "visible alternative timescapes [that are] enriching our temporal imaginings" (Puig de la Bellacasa 2015: 707). As such, even if it is against the very aim of them as campaign narratives that should impel urgent action, they are not merely about care as a progressive enterprise. They are rather about the coexistence of different temporalities folded together, and differently folded through acts of witnessing (by the storytellers and by the audience, including myself).

Risk, fear and relational care

Risk figures in the campaign. For example, in Cecilia's narrative it was the doctor at the hospital who told Cecilia after she had received the cancer diagnosis that surgery is usually enough and, therefore implicitly, that the uterus *most often* does not have to be removed and that you *most often* do not die. But then, she thought "this word, 'usually' – what does it mean?" This is also related to the excerpt already mentioned from her story included in the campaign image: "What if it has spread through the body? What if nothing can be done?" Then the narrative tells us that, in fact, her uterus had to be removed. This is a narrative that articulates risk; risk of tumors growing too quickly, too uncontrollably, and of potentially being one of the few where the uterus has to be removed – or who may die.

The risk described in Cecilia's narrative relates to fear as a matter of anticipating an uncertain future. It is about invoking the *it may happen*, or *it won't*. It is about cancer death as always, already here. As an affective, temporal risk state, we are living it. In depicting life as contingent and death as an always, already present trope and reality, it is, once again, the possible (the *it may happen*) that is played on. This can be understood as an affective risk state; it is the potential *possibility* of cancer that generates feelings. It is *cancer anticipated* that generates feelings.

As is familiar, this is often what "being at risk" is framed as being about in our contemporary society (Lupton 1999). It is about being positioned in a state where you know that something can happen to you, but you do not know if it actually is going to happen. However, Cecilia's narrative is not merely about her being at risk. The risk talked about in the narrative is a risk located within relationships to others. In Cecilia's narrative it is, for instance, written: "When I got sick, I think it was worse for my husband". In this way, it is a story about risk ("what if it has spread in the body?"), but it is a risk linked to care for loved ones.

Risk as an affective state is often directed toward the individual in public health campaigns (Gagnon et al. 2010; Lupton 2013; Porroche-Escudero 2014). The second "I love me" campaign differs from this. The focus in Cecilia's narrative – and in other examples of the narratives where risk is evident – is a risk embedded in loving and caring relationships. This differs from other campaigns using death rates and the statistical probability of cervical cancer diagnosis to make people scared, and campaigns that are staging a fear first and foremost for one's *own* life.

In the second "I love me" campaign, the narratives are full of affective memories of loss, pain, suppression of feelings, of death. If the campaign is about fear, it is a collective, relational fear that encourages people to care for each other. For example, in Ida's narrative, one finds fear, but it is also about love, expressed through the fear of her mother being taken away from her. "It felt better to take a step back and to leave the ones

I loved, instead of having them taken away from me". Ida's desire to not relate (but failing to do so), denotes a care that circulates around both fear as an unhappy feeling and love as a happy one. It is a care that includes specific versions of fear and love embedded in close relationships, and transformed due to the cancer.

As Mol (2008: 67–68) points out, fear-based public health campaigns often start with individuals and aggregate to the collective level. The "I love me" campaign, instead, entails "a logic of care" (Mol 2008) as its starting point is relational feelings and experiences are embedded in caring relationships. These relationships include being at risk as an affective state of, amongst others, fear and love. Importantly, this means that "being at risk" is not simply articulated as individual risk.

Affective life-changing times

Kristina's narrative is full of anxiety, pain, despair and fear. It invokes a sense of desperation. It starts with a description of the tumor in her body and the cancer:

> "It's cancer", the doctor said. I asked if it was good or bad. But cancer is cancer, a tumor is a tumor – and it's evil [...] Pains. Bleedings. Cramps. Due to it all, I felt extremely bad psychically. Help me now!

In describing pain and desperation, Kristina's narrative is affectively dense. It invokes the idea of a tumor that attacks the body, from *within* (like an evil stranger). The body is endangered by the tumor, something that leads to a desperate call for help. The narrative re-presents a subject that is not in control. Instead an evil and uncontrollable tumor threatens to take over the whole body. The main part of Kristina's narrative is, in this way, focused on the "darker sides" of the cancer experience. Nevertheless, her narrative ends in a hopeful tone. She writes, "My husband and I are closer now. We're happy that we have each other, and are opening up toward each other more. We don't want to lose one another".

A similar focus on how relationships with relatives have changed due to cancer appears in several of the narratives in the "I love me" campaign. The cancer changed the women's and relatives' approach to life and to relationships with loved ones. They do not take life or their relatives for granted any longer. "You must live now!" is stated in Cecilia's narrative. In these examples, emphasis is placed on how the cancer brought the patient and the relatives closer to each other. Because of the cancer diagnosis, they understand how unpredictable life is.

The emphasis on transformed relationships with relatives is an important theme also in Lukas's story:

> I take nothing for granted today, despite the fact that my mom defeated the cancer. And I have a new perspective toward my parents. We are not just family members – we are life companions.

The narrative tells us about a transformed way of relating to life and to close ones. The cancer is described as having the capacity to change relationships between humans: from being family members to being life companions. Thus cancer is not depicted merely as something relational but also as something that generates (changed) relationships. This is echoed in Kajsa's narrative, which ends: "I have a wonderful boyfriend and I have started to be able to tentatively trust life again". If the cancer unsettled her trust in life, her relationship with her boyfriend is said to help her trust again. These narratives are about how life is transformed when cancer strikes and about how relationships to loved ones become more intimate.

With their focus on survival, these narratives invoke hope. This is common in "cancer stories" as a circulating phenomenon. Narratives about cancer survival (like that of Kristina) often include a focus on how the former cancer patient is now closer to her/his relatives. A simple googling of "cancer changed my life" gives thousands of hits, many from relatives telling us about how their life was changed and how they no longer take their close ones for granted. "Cancer stories of hope" generates even more. Many of these are very similar to Kristina's. Jain (2013: 31) criticizes this kind of survival story for shifting the emphasis away from cancer commonalities (and cancer as disease, fear, pain) and toward the "fighting individual" through a "triumphant ideal of the human spirit". In this sense, such survival stories are examples of a "classic triumph-over-tragedy cancer narrative" (Stacey 1997: 21). However, in the "I love me" narratives, the focus is not on individual survival and triumph. Instead, where the narratives invoke hope and survival, they do so through the emphasis on caring relationships.

Lukas's, Kajsa's and Kristina's narratives encourage us to not take anything for granted. However, "anything" here is a specific anything: do not take for granted that you or your loves ones will remain cancer-free in the future. This again is made possible by invoking cancer as always, already here. The transformed relationships described in the narratives articulate care as related to being life companions, as *really* having your close ones in your life, as sharing life with each other, instead of simply assuming that nothing will happen (like cancer). Cancer is articulated as generating more intimate and caring relationships and a transformed way of handling life.

Describing life as changed also implies a transformed way of anticipating the future. "Anything can happen in the future. You need to live now", says Cecilia. She now approaches her life and possible future in a different way, and anticipates possible scenarios in the future by living life differently today. In this sense, "anticipation is not just a reaction, but a way of actively orienting oneself temporally" (Adams et al. 2009: 247). This is reminiscent of the narratives about changed relationships to relatives. By having experienced how (possible) death can appear in their lives, they no longer take life for granted. Life is articulated as contingent and uncertain, and death not a something abstract and foreign but something that can happen here and now. This does not only link life, death and time, but also feelings. For example, in Kajsa's story the cancer tumor is affectively described – pains, evil tumors, "help me now!" – but ends with hope and changed relationships with loved ones. Loving relationships, fear of death and hope for life are articulated as linked.

Another example of how temporality and feelings are linked in the campaign is in how reproduction and future children are discussed. Not to be able to have biological children is one of the things brought up most often in the narratives. Kristina states that "the worst thing of all is that I cannot have biological children". Her partner Erik talks about the same issue but in slightly different terms:

> The most important thing for me is that I have Kristina. That I can live with her for the rest of my life, have her by my side. Sure, we want to have kids, and we will try to make that happen in one way or another. But Kristina met a mom who put it in such a good way: You don't need to carry them here (laying her hand on the stomach) to carry them here (laying her hand on the heart).

In Kristina's and Erik's narratives, their wish for children is emphasized. In Kristina's narrative it is described as being about the desire to have biological children, something that is impossible as her cervix has been removed. In Erik's story, in contrast, the possibility of having children through other means is stressed. Despite this difference, it is striking that future children take up such a big part of their narratives. Imagining life differently due to the cancer diagnosis is here connected to having to relate to future children differently. Reorienting one's life temporally ("you need to live now!") is performed as including a need for reorienting one's desires for biological children. In these narratives, love includes the vision of an ideal future that includes a heterosexual couple with biological children.

Cecilia's and Ylva's narratives are about already having children or about still being able to have biological children. From Cecilia's narrative: "I had my kids, and did not want to have any more. [...] But I realize how lucky I am; the fact that it was "only" the cervix and uterus that were affected. And I'm so happy that I have my kids. I know that others are not as lucky." In a similar way to the previous narratives about children, Cecilia's narrative foregrounds children as an important part of her life. The difference here is that she already had children when she was diagnosed and her uterus was removed. The focus on luck is also present in Ylva's emphasis on children:

> The operation went fine. And I will be able to have kids! But it will include risks. I cannot give birth vaginally; instead it needs to be a caesarian. The risk of a premature birth and miscarriage is a third higher. But it can work out. Most likely it can work out! I have had a great deal of luck in the midst of all the misfortunes.

The possibility of being able to have kids outweighs the possible risks, something which exemplifies how strongly the narratives present children as an essential part of a desirable future. In these narratives, an ideal future is invoked that is negotiated and transformed. It includes a reorientation of present desires and wishes for children and, as part of that, transforms people's hopes for the future.

The presence of articulations of future children in the narratives serves to reproduce a "heteronormative timeline" (Taylor 2010: 894) that links a desirable future with children. Future children are here made present as an affective investment that the narratives ask the audience to care for. Children figure as a promissory, hopeful and "obligatory token of futurity" (Edelman 2004: 12). In this sense, a vision of "reproductive futurism" (ibid.: 131) is reproduced, and a "happy future" trope gets closely linked to reproduction and children.

In both Cecilia's and Ylva's narratives, luck is staged as crucial. As Ylva says: "I have had a great deal of luck in the midst of all the misfortunes". Luck invokes uncertainty and unpredictability: you may be lucky, or you might not. This turns the future into something uncertain. In this sense, Cecilia's phrasing "I know that others are not as lucky" can be read to tell us that it did not happen to Cecilia but it can happen to you. Do not rely on luck, get vaccinated. That is something controllable. Luck is not. "Anything can happen in the future" and it is "an unfair cancer". *Unfair*, as already mentioned in connection to Lukas's narrative, denotes that cervical cancer not only has to do with how one has lived one's life, but also about things out of one's control. In this way, *unfair* and *luck* denote cervical cancer as something uncontrollable for the individual.

Luck is imagined as an uncertain affective state, as something you cannot control but which *may* be yours. If you are not lucky, you might not be able to have children. This articulates the message that you cannot control luck, but you can decide not to build your life on luck if you decide to get vaccinated. This brings the focus away from individual morality. The focus is not on Cecilia or Ylva as having done something to cause the cancer – luck puts the focus away from individual agency (from a moralization of health). At the same time, being part of a campaign for HPV vaccination, luck is implicitly articulated as the bad option and HPV vaccination as the better one based on the fact that it promises an alternative to holding on to luck and uncontrollability. As is often the case in HPV vaccination campaigns, this reduces uncertainties about HPV vaccination by staging it as the "right tool" against cervical cancer.

Conclusions: affective and temporal cancer narratives

I have focused on affective relationalities as re-presented in the campaign narratives and as articulated in the meeting between the narratives and the audience. I have attended to this as matters of care re-presented in, and invoked by, the campaign narratives *and* by engaging a retelling of these narratives as a matter of care in itself. In doing so, I have discussed how people are asked to care for the storytellers' narratives, and how it is possible to carefully respond to this encouragement to care.

I have discussed a range of matters of care re-presented in, and invoked by, the campaign. I have stressed that death and disease are visually absent in the photos, and how the bodies of the storytellers are re-presented as healthy, unmarked bodies. Through their intense gaze, the storytellers are re-presented as in control of themselves and their story. I explained this as a specific matter of care: viewers are invited to care for *their* stories. Moreover, I discussed how cervical cancer in the narratives is re-presented as a disease that is a matter of care not only for women, but for relatives, as well. For example, I argued that *falling apart* and *breaking down* were invoked as affective states in some of the narratives (including my own), and how this sometimes was linked to the trope of a rational and protective male subjectivity, and a gendered matter of care. Additionally, I problematized how care figures as a mode of identification in the campaign as this encourages people to care for the storytellers' narratives on the basis of them seeing themselves (or, rather, their potential future self) in the storytellers. In these matters of care, subjects that care, or that are being cared for, are articulated and envisioned.

As is the case throughout my study, several of the matters of care represented and invoked in the campaign have to do with temporalities of care. One such matter of care is the idea that we all have a relationship to cancer. In the campaign, people are addressed through an always, already presence of cancer. Therefore, cancer as an affective state, I emphasized, is about possibility and potentiality evoked by the circulation of cancer stories and realities. It is about invoking the *it may happen,* or *it won't.* I discussed this partly as being about a future-oriented anticipatory immediacy in how this is drawn upon in the narratives to encourage people to get vaccinated now. A related future-oriented temporality was the care for future children circulating in the narratives and, connected to that, the hopefulness invoked in the narratives about transformed relationships to loved ones, and to reorientations of current and future life.

Yet I have argued that the narratives include also other temporalities of care. I have emphasized that they fold past, present and future time in how they re-present caring relationships. Thus, they include coexisting temporalities of care.

Finally, by attending to the range of feelings re-presented in, and evoked by, the campaign, I problematized my own "gut reaction" about it being first and foremost about scaremongering. The narratives are, indeed, full of affective memories of grief, love, loss, anxieties, pain, hope, shame and fear. By attending to how these affective registers are blended in the narratives, I have been able to capture the multivalence of affectivity the campaign inhabits and invokes. Moreover, in paying close attention to how different feelings are evoked when I read the narratives, I emphasized the complex and non-unidirectional relationship between the affective relationalities re-presented in the narratives and the ones provoked or generated by them.

Epilogue: responsible re-storytelling

Staying with my discomfort over my "gut reaction" toward the campaign, and by focusing on affective relationalities as engagement and commitment, have helped me think differently about the campaign and, hopefully, do some useful critiquing through responsible re-storytelling. I have interwoven narratives from the campaign with my own cancer-related story to show how the campaign includes a wide range of feelings of care and invokes affective states. In this way, I emphasized how affective memories about my father's cancer were invoked when I read and viewed the campaign narratives. I have aimed to stay true to the narratives and the mediated affective experiences they invoke, as I think this is one important way to understand what cancer is made into in contemporary society. It provides insights concerning "how cancer becomes us" (Jain 2013). Moreover, taking the narratives seriously by discussing them, as a matter of care invoking responsibility and ethics – and as inhabiting matters of care – is a way of responding to the narratives. This practice of re-storytelling, hopefully, can help foster a "learning to become affected" (Schrader 2015: 684) as a careful, and caring, engagement.

At the same time, the narratives told here are not just *any* cancer narratives. They are part of a campaign about HPV vaccination designed by a county council. Telling cancer narratives as part of an HPV vaccination campaign reproduces reductionist ideas of Gardasil as a vaccine against cervical cancer, something which links the campaign's narratives to other narratives about progressive, successful biomedicine finding a cure against cancer. This clearly shows the non-innocence of the politics of care in technoscientific governance. In a context of HPV vaccination, such narratives make absent not only boys as possible HPV vaccine users but also the fact that current HPV vaccines are estimated to protect only against 70 percent of the HPV types that may cause cervical cancer (and that *estimate* includes uncertainty). Learning from Jain (2013), this focus on success and hope (and, indeed, HPV vaccines as cures against cervical cancer) makes it possible for us to be ignorant of the intimate and paradoxical place cancer has in current society. It is important to be attentive to how the second "I love me" campaign focuses on caring and loving relationships can make absent the multiple public health politics involved in HPV vaccination practice. For example, and as I will discuss further in Chapter 9, the emphasis in the second "I love me" campaign on caring and loving relationships between relatives makes absent broader questions of herd immunity and population care, and it easily reproduces an imaginary of care as something inherently unproblematic.

Being part of an HPV vaccination campaign, the cancer narratives carry with them certain ethico-politics of care. It is impossible to anticipate generated feelings – not least in a context such as cancer with all its circulating, affective stories, memories, futures, and realities. Therefore, the stakes are high in designing a campaign such as this one. If this can be done in a responsible way or not is something to live with, not to answer once and for all. By staying with the multiplicity and generative troubles and promises (for good and bad) of the campaign, I argue that it is possible to respond to, and engage with, the campaign narratives without using once-and-for-all solutions and ready-made explanations. The focus on care as a matter of responsibility will be further discussed in Chapter 9.

9. Communicating Death and Disease with Care

"Facts did not convince the target group", my interviewee Klara said. Therefore, she continued, we needed to show that "people actually die" from cervical cancer, and that whole families get strongly affected by the disease. Moving quite drastically from a message "as positive as possible" to cancer narratives about death and disease, Bredland County Council hoped that depicting and addressing people's care for their relatives as evident in situations of disease and death would make young women and girls decide to get vaccinated. As shown in Chapter 8, this made central versions of care other than care as self-love and happiness (as was prominent in the vision of the first "I love me" campaign). Even if love was invoked in this context as well, it was a different form of love since it became connected to affective experiences of disease and (possible) death.

For my interviewees the focus on death and disease in the second "I love me" campaign brought about questions of responsibility and accountability. The campaign's message was by my interviewees stated as "unusually strong", and they argued that they therefore needed to communicate the campaign message responsibly. They stressed that the county council has to work with values of care, and that this focus on care would enable them to responsibly communicate death and disease to the concerned public. Care was by them sometimes articulated as a "clear-cut solution" to problems with communicating death and disease. This chapter centers on how public accountability, responsibility and care were linked in the interviews. I especially focus on my interviewees' articulations of the need to communicate death and disease in a caring manner.

I draw upon material from my conversations about the second "I love me" campaign with my interviewees in Bredland County Council. I mainly discuss how Klara, Head of Communications, and the communicator, Helena, talked about the campaign. The epidemiologist Emma and the administrator Linnea were not involved in the work at Bredland County Council with the second campaign. However, as I asked Emma about what she thought of the campaign (I did the interview with Linnea before the second campaign was launched and before I found out about it), I will also bring in some of Emma's comments on it.

Several feminist STS scholars have discussed links between care, accountability and responsibility. Jerak-Zuiderent (2015) shows how situated care and accountability might co-emerge from within the practices of health care. In a related vein, Vicky Singleton (2012) shows how accountability is done in local practices of collective care and responsibility toward a collective. Yet another example is how Kenney (2015: 750) suggests "that 'accountability' could be a useful name for the politics of knowing, caring for and building worlds in STS". This, she proposes, makes it possible for feminist STS scholars to become more thickly connected to the matters of care engaged with as world-making practices.

These scholars shed light on how practices of being accountable for, and caring for, subjects, collectives or worlds involve specificities and partialities. Moreover, and importantly for this chapter, they highlight that accountability as an expert practice may include moments of responsible and situated "accountability *with* care" (Jerak-Zuiderent 2015: 429, emphasis in original) generated *from within* the studied practices. In doing so, they convincingly show how "care is not a new dimension feminists are bringing to technoscience but rather an already circulating [...] force in our worlds" (Murphy 2015: 731), and that these politics of care can simultaneously close down, and open up, for practices of caring well.

In the interviews about the second "I love me" campaign, moments of closing down, and opening up, for accountability *with* care coexisted. Importantly, and even if care was sometimes articulated as a "clear-cut solution" to problems with communicating death and disease, there were many moments where a clear-cut status of care was put into question. *How* care can be a solution for accountability problems was not a given.

I start by discussing authenticity as a vital matter of care invoked in the interviews, and as enabling responsible communication. This is followed by a discussion of how relational care and death were articulated, and delimited. I then focus on the politics of how the notion of the herd as a care for others was made absent in favor of a care for the family. The chapter ends with a discussion of what the matters of care brought up in the interviews might say about links between care and accountability.

Invoking authenticity, "demediating" the medium

At first glance, the communicator, Helena, and the Head of Communications, Klara, explained the campaign message as something quite straightforward: you should get emotionally affected by the stories and therefore decide to get vaccinated or decide to tell your close one to do so. (At the same time, and as will be clear in this chapter, this does not mean that they thought it was easy to convince people to get vaccinated.) I asked them why they decided to design a new campaign so different from the first "I love me" campaign. Helena answered:

> HELENA: We needed to be more emotional and more serious.
>
> LISA: And that you thought since …?
>
> HELENA: To increase attention, to show that this is serious and in this case, cervical cancer is actually something you can die from. And if you don't, it is a really tough treatment that you need to go through. And, therefore, we wanted to elucidate that and, thus, we decided to work with storytelling, as we call it. That is, you let real people tell their stories. Both as an afflicted [woman] and as a relative.

Helena stresses that the use of storytellers' cervical cancer experiences in the campaign enables the message that cervical cancer is something serious, something you can die from. This is based on an idea that, if the stories are real, then the seriousness of the disease is real. Possible future death is something real. Later in the same interview, when I brought up the emphasis on real stories again, she stressed that it was important for the county council that the stories were real because otherwise it "would not feel authentic". The stories should be "collected from reality" and "shouldn't be edited in any way". Klara, too, emphasized the significance of the realness of the stories and of the people. As she put it, it "becomes more authentic". Thus Helena and Klara both stressed that the "I love me" stories told about authentic memories of cancer, and by extension, authentic memories of feelings. To them, reality and authenticity were conveyed when they let real people talk about their own experiences. This would make it *feel* real. One authenticity is translated into another: real relations and experiences are envisioned to generate real feelings and, therefore to affect and move people (to get vaccinated).

This implies that real and affective stories *merely collected from reality* will make people understand the realness of cervical cancer. This is a specific version of reality: stories are simply extracted from an already existing reality, in contrast to realities as being made or remade. It includes an idea of a reality where the method of storytelling through text and photographs does not mediate – that is, transform, displace, translate – cancer experiences. This idea makes the campaign medium used absent; it is imagined that re-presenting cancer experiences in a campaign tells us about unmediated – instead of specific and partial – experiences. A "demediation of the medium" (Kember and Zylinska 2012: 158) is occurring.

Klara and Helena mentioned that "unreal stories" would be made, for instance, if they were to hire a model that had the looks but did not have the real experience of being a cervical cancer survivor, or the relative of one. Helena said that many people assumed that they had used models and fabricated the stories, but:

> [It's] really important for us, that it's real stories. I must say that I thought it was really interesting that we got so many questions about if it was for real or not. The majority probably assumed it wasn't. But it feels really strange to hire a model and then fabricate together a history.

Hiring a model and fabricating a story would not illuminate the realness of cervical cancer. Fictive stories would not make the stories *feel* real and would therefore not produce a strong affective reaction. Klara stated that models would "probably be a bit good-looking, and less varied in style". On the contrary, using people who tell their own stories would allow for "an actual mixture of how relatives [and afflicted women] look". Variation in style thus translates into real people telling real stories.

A vision of *authenticity* is crucial in how Klara and Helena discuss the cancer stories. As a follow-up question to Helena's statement about models, I asked her what would have been different with having models instead of real people. I wanted her to elaborate more on why it was important to have realness and authenticity. In her words: "If it wasn't for real ... Then it wouldn't be for real. Then you can't embrace the message". The "real" is emphasized as important due to it being real: if it is not real, then it is not real. And if it is not real, then people cannot embrace the message. By giving the real a self-explanatory weight, it is assumed to have the capacity to generate affects – and thereby to change behaviors and anticipate cervical cancer. The idea is that if stories are real they move people, affect people, change people. Thus, a vision of authenticity as unmediated cancer experiences is staged and is imagined to have the capacity to convey a *feeling of* realness and authenticity, something that made up stories would not. In other words, authenticity figured as a matter of care as it was imagined that it would get people affected.

In Chapter 5, I discussed a *view from someone*. Such a view is also present here, yet in a different form. If the view from someone there was equalized with *facts from someone*, in the second "I love me" campaign *stories from someone* were hoped to convince people. This is interesting as experiential stories as a basis for vaccination decisions are often problematized in a vaccination landscape that privileges science-as-epidemiology. As discussed in Chapter 5, stories were strongly problematized as a basis for vaccination decisions by my interviewees at Mittland County Council. In their opinion, stories needed to be responded to with facts. The stories talked about in that context were stories about side effects that were understood as "merely stories", that is stories without a reality referent. Different from that, in the context of the second "I love me" campaign, stories were emphasized as real, authentic testimonies. In the different contexts, stories were valued strikingly differently: as rumor and myths, or as authentic real ones.

Even if a vision of authentic stories is not about playing the god-trick as these stories are *from someone*, it inhabits a "representationalist" imaginary in how it depends on a trope of an external reality that can be displaced without getting transformed. The stories are envisioned to carry the weight of facts since they promise authenticity.

Several times during the interviews, Klara and Helena brought up that people were critical toward the focus on, as Helena formulated it, the "black and death". From my and Helena's conversation:

> LISA: But this focus on the death and the dark black, what response have you received on that?
>
> HELENA: It is different from the old one, it's another approach. For the good and the bad. Then there's many who think this is something awful to talk about [in a campaign], to point at death and misery. It's real people that are telling their story [...]

Helena made the point that many think it is awful in a campaign to show "death and misery". After having said that, however, she mentions that the storytellers are real people telling their story and that the HPV vaccination prevents girls from being afflicted with cervical cancer. In other words, communicating death as part of the campaign was understood as responsible as long it derived from real experiences. It was articulated as a matter of care when it was being linked to authenticity.

At the same time, it was not formulated as an easy demarcation. It is not a given that communicating death can be a matter of care. In an earlier interview Helena had emphasized that "of course, some people think it's dark as night and it's a classic balancing act in health promotion work, where you, you know, should draw the boundary". This can be interpreted as being about the difficulties in this context in drawing a boundary between what care is and what it is not. It is a balancing act to communicate a dark and black message that also needs to communicate care. Letting people tell their stories about death (and doing so with a focus on their relationships to close ones) is for Helena hopefully to draw the boundary at the "right" place.

Authenticity as public accountability

Using real people instead of models was not only linked to an idea of telling stories about real experiences. It was perhaps more importantly connected to articulations of the county council's responsibilities to the concerned public. Both Klara and Helena emphasized that the second "I love me" campaign differed from how public health communication was most often designed in Sweden. There is a prevailing idea that health communication "should be first and foremost fact-based and you should not look commercial", Klara said. This differs from how the second "I love me" campaign centers on feelings. As Klara indicated, this makes it close to commercial advertising in its design, as commercial advertising often plays on feelings to address consumers (see e.g. Sturken and Cartwright 2009). Importantly, however, authenticity was stressed by Helena and Klara as the reason for why it was *not* like commercial advertising. In Klara's words:

> Of course a commercial actor can also have [...] non-models but they usually don't. But we can. And our trademark is to be more in accordance with that than models. So that's not that weird really, for us it's both logical and good to pick those that have been through this [...] [I]t works better, I believe, for us as a public organization ... There's an engagement behind the images. These images are not there because they got a specific fee paid by the hour, but because they care for this issue [...] There's something with that we think is important.

Klara emphasized that letting people tell their stories speaks to "the trademark" of the county council. The county council's trademark does not go together with hiring models that are participating in the campaign design because they get a "fee paid by the hour". Later in the interview she talked about this as being about "the values [the county council] wants to communicate" and that "these are values that are warm and caring". Care was invoked as being in conflict with financial compensation (such as getting paid by the hour) as it was assumed that when people really care about an issue they do not need money for participating. Warm care was articulated as in conflict with cold money. To really care was envisioned as being outside of financial matters. What is more, money would prohibit, or distort, care.

Real cancer experience is translated into the storytellers caring about the issue. In Klara's reasoning, this is the reason why they care, and why the message is one about care. Thus, that they care is envisioned as closely linked to authenticity. They care as they are authentic witnesses. Here, authenticity becomes a matter of care due to its *separation from* financial compensation and from commercial actors (pharmaceutical companies).

STS scholars have problematized the idea that care and money have to be in tension with each other. Care and money today often presuppose each other; different "economies of care" exist (see Harbers 2010; Eidenskog 2015). What these studies stress is that the inclusion of money does not have to mean a loss, or distortion, of care. Instead, "economies of care" enable specific versions of care. Even if the storytellers were not paid for telling their stories, the campaign by no means exists outside of financial matters, public authorities like county councils are part of broader welfare economies. And as the catch-up vaccination is part of the care choice system, the county council's work is entangled with a marketization of care. However, in the interview with Klara, the second "I love me" campaign was articulated as about care since it did not include models financially compensated.

Klara stated that non-models work better for them, as the council is a public organization; this choice is more in line with their trademark and values. That the county council needed to design a campaign that was in accordance with the county council's values has to do with articulations of public accountability and public trustworthiness. Accountability as a matter of care was invoked as in line with how the county council needed to communicate information. That is, it did not only have to do with what they believed would increase the vaccination coverage, it also had to do with the county council's values and responsibilities to the public concerned. That the storytellers "really cared" about cervical cancer and HPV vaccination – and that they therefore were not financially compensated – gets linked to a question of how a county council needs to take responsibility for its public communication. The county council needed to be accountable to the public concerned – and therefore hold on to their values and their trademark – and this was evoked through a vision of authenticity as (warm) care, and a distancing from (cold) money and commercial actors.

It was envisioned that people would not be the same people if they received financial compensation. Their, and thereby also the county council's, trustworthiness would decrease. The boundary between the county council, as a public authority, and commercial actors, such as pharmaceutical companies, would risk becoming blurred. The county council's trademark would be at risk of becoming questioned, as it would not act in line with the values of a public authority, values that are distinct from the values of commercial actors. According to this reasoning, what holds the county council on the right side of the boundary is the fact that the storytellers are real people. It remains on the side that depicts and communicates care to the public in a responsible manner.

Listening to girls and young women to enable care

Klara and Helena emphasized that the focus on disease and death in the campaign was the result of the county council working with, and listening to, the target group (which they did through the interview evaluation discussed in Chapter 7). With a reference to the campaign as being black-and-white, Klara, for example, stated that "[i]t's not that much due to someone in the project group or [design] agency liking black". The campaign was dark, black and about death since girls and young women had told them that a darker tone was needed to make people understand the seriousness of cervical cancer. Klara said:

> [In the campaign we] go from "think about yourself" and "love yourself" to actually connecting the vaccination with death. How could we do this? And I know that we had discussions when we got drafts for the campaign. Internally many reacted to this. But we continued as we chose to develop the campaign together with the target group. And if the target group says that "we think this will work" [...], then we have done what we can.

By listening to the target group, and developing the campaign together with the concerned public, they "have done what they can". This is a specific matter of care that I have discussed throughout this study: it is emphasized as fundamental to listening to girls and young women addressed by the vaccination program. What is more, it is to listen to them in a very specific setting of public participation where a few girls and young women get to re-present an imagined "target group population" existing "out there". This makes absent how the evaluation is a device which enacts a specific version of what it means to listen to the girls. Its partialities and transformative capacities are made absent. When the evaluation shows that girls and young women "out there" think a death focus is needed it is assumed that such an approach can be caring and responsible. Listening to the girls and young women through an evaluation translates into collecting opinions "out there", something that, in turn, translates to be caring and taking responsibility toward the concerned public.

By asking "how could we do this?", Klara implicitly referred to an idea that it would be problematic for a county council (or any other public organization) to work with death-focused communication. As she stated, this is not a common communication strategy for public organizations. Asking "how could we do this?" was not expressed as a mode of self-reflection but rather as a rhetorical question. As such, it sheds light on the lingering presence of the very idea that it would be problematic to communicate this kind of information. Klara's rhetorical question hints at a tension between communicating death as irresponsible or communicating death as caring for girls and young women. However, for Klara this tension was "solved" by reaching out to the target group and taking seriously what they thought could work.

This was not simply seen to be about finding the most efficient way to increase vaccination coverage. Klara rather emphasized that it was important to collaborate closely with the girls and young women, as this is an "obligation" they have as a public agency. She stated:

> I think it is a formal obligation, too. If the county council wants to reach out, there is an obligation to reach all citizens. And then we need to be where they are. Then we can't think that we choose the cheapest way [...] Or that we decide on what is easiest for us. Instead, we need to be where the target group is. This is both a formal obligation and an approach we have.

In this quote, being where girls and young women are is presented as an obligation and an approach that they must have as a public agency. This is important in how it complicates an assumption that their communication strategy would only be about the end justifying the means (that they are trying to find the most efficient way to increase the vaccination coverage). Instead, it was articulated to be about working *from within* a governmental regulatory frame stating that they as a public agency *need to* listen to concerned publics. Reminiscent of how I have already discussed financial matters, the above quote articulates a tension between care and finance. The county council cannot decide on the cheapest option; it needs to listen to the concerned public. The county council's public obligations were also present in how Klara argued that they needed to work with values of care. It was not her personal opinion; it was about them being a public agency.

Here, links between accountability, standardization and care are articulated, such as how policies and legislations condition and enable care work, and how public participation is involved in this work. I see it as an example of accountability *with* care where a general framework of governmental obligations is adjusted, or "tinkered with", to meet the requirement of being accountable to *this specific* public.

Affective relations of care

As discussed in Chapter 8, the presence of relatives is important in the campaign. My interviewees emphasized this as being a crucial reason for why the campaign was about care and why it therefore was a responsible endeavor. Klara and Helena discussed this inclusion of relatives as a way to reach out to more people. It was not merely about reaching girls and young women. It was about reaching them *through* their relatives. In Klara's words:

> [I]t's a way of covering a bigger part of the target group. To work not only with traditional witnesses; "I had cervical cancer, now I have survived. Get vaccinated, you too". Instead, you reach the target group from several different directions [...] [T]his one [Klara points at one of the storytellers re-presented as a campaign image] wants [you] to place yourself in the situation of [having] a deceased sister or cousin. It's [another] way of reaching the target group.

Klara emphasizes that reaching young women and their relatives is about identification; it is about enabling relatives of young women and girls to identify with other relatives, relatives who have cancer experiences. The storytellers, in Klara's words, encourage people to place themselves – by reading and viewing their stories – in their situation. It is thus an invitation to put oneself in someone else's shoes. It is believed that people looking at the campaign will grasp what the storytellers *really* feel and therefore realize that "I don't want to feel like that".

Through such a mode of identification, feelings are envisioned as unmediated and immediate. The idea of authenticity *as* care is invoked as being enabled through identification with unmediated feelings. Such "caring for" as identification comes with an ideal of sameness and, as such, it is a conditioned, and often exclusionary, form of care. In Klara's words this vision of identification as sameness gets closely linked to an ideal of authenticity.

Helena explicitly related the county council's hope to reach the target group from other directions to a wish to communicate how it feels for the people affected by a cervical cancer diagnosis. In her words:

> We thought that we wanted to get at what it may mean for these people that are affected. How do you as a person get affected? How are you affected when you get this diagnosis? And also, how are you affected as a relative? You know, it's the whole family that is affected.

In this quote, Helena emphasizes the family as the unit being affected by a cervical cancer diagnosis. Cervical cancer is invoked as a matter of care for the whole family. Expanding the HPV vaccination audience from girls and young women to the family challenges predominant HPV vaccination campaign narratives which center on girls or women (Braun and Phoun 2010; Vardeman-Winter 2012; Charles 2013, 2014; Davies and Burns 2014; Burns and Davies 2015). At the same time, however, with the inclusion of the family as the campaign focus comes an exclusion of other possible people and constellations. This is based on a family-centered trope assuming the family as the main "care unit" in adult life (see e.g. Borchorst and Siim 2002).

However, when the epidemiologist Emma talked about the second campaign, she did not only mention the family. She also included friends as the close ones addressed by the campaign. "You bring up the family, as a friend or a sister or … That's also something one cares about … somebody else one cares about that it can affect," she said. Through these words, friends are included as caring for a close one who is/was a cancer patient. I do not think Emma reflected upon what she said; in the context it did not seem like a conscious decision to include friends as well. As Emma had not been involved in the work with the second "I love me" campaign, perhaps she thought that some of the narratives in the campaign did include friends as well. However, her formulation still opens up the possibility of including friends as caring subjects, and as such it is a moment that problematizes a family-oriented normative ideal about who gets included in a group of "close ones" or "loved ones".

Men are often absent in the context of HPV vaccination, so when reading the narratives in the campaign, I was especially intrigued by the inclusion of men. When I asked Helena about this she stressed that "it's good to have the guys too [in the campaign] as everyone can see and react and think 'but oh my god, has my wife gotten vaccinated?!'" She continued by stressing that they can "really go home and ask" and "perhaps push [their partner] a bit". In this way, HPV vaccination communication was articulated as a phenomenon that concerned both men and women. Men were considered important; they could support, encourage, protect. At the same time, being positioned as "helping out", young men were envisioned by Helena through a gendered divide: cervical cancer was a women's disease that men emotionally could assist with.

How men here are envisioned includes an assumption of male partners wanting to protect their female partners. When care is made present as relationships with close ones, it comes with gendered assumptions about *how* men care. They are invoked as caring for their partners through a desire to protect. This is a gendered trope, which identifies females (and children) as in need of male protection. Even if this mode of address is different from an individually focused self-care, the message still articulates public health communication as a case of *gendered* care.

It is worth mentioning that Helena did not always discuss the inclusion of males by invoking gendered tropes. For example, in another interview she emphasized that it is good to include boys as they are also emotionally affected by cervical cancer. "It's not easy to see the one you live with, the one you love, becoming sick and having to go through a tough treatment," she said. In this passage, males are made present as people who experience cervical cancer themselves, and not merely as people who are "helping out" or are protecting their female partners. Instead, they are envisioned as possible addressees that are a part of cervical cancer worlds. Without it necessarily being along a gendered divide they are assumed to care for their partner, and get affected by her disease. This opens up an alternative articulation of how males may care, and be affected, in the context of cervical cancer and health campaigns, one that is not necessarily about a gendering of care.

Gender was also brought up later on in the same interview with Helena when she stated:

> [W]e have seen actually that [...] the guys' screens did get less attention. Perhaps there's still something feminine [about] cervical cancer ... [silence] That you, kind of, like, "oh God, look there, she, and think if I?" ... Instead of connecting to a guy. [silence] I don't know.

Helena states that the county council has seen that the males' screens received less attention. She refers to an evaluation that the council did of the second “I love me” campaign. Similar to the evaluation discussed earlier in this chapter and in Chapter 7, this evaluation was invoked as a device that re-presented the opinions of the concerned public. Using this device, Helena suggested that perhaps it was better to address women directly rather than through relatives; it was perhaps better to have an individual, female-centered, focus (as is often the case with HPV vaccination campaigns).

When stating this, Helena hesitated. Being silent in between the sentences, and adding a “perhaps” before “something feminine” and by ending with “I don’t know”, her answer included uncertainties and precautionary hesitations. Stating that cervical cancer is perhaps (but “I don’t know”) something feminine is not the same as saying that it is. Instead, a degree of speculation was present: it may be like this, or it may not. This differs from how she previously talked about the inclusion of guys as something good. The evaluation was invoked as a device that enabled speculation rather than gave a distinct, clear-cut answer. The evaluation was drawn on to enable a problematization of *how* the county council cared, and re-presented care.

As I also discussed in Chapter 5, attending to hesitations and silences can be a matter of care. It may trouble a vision of knowledge as progressive and stable, and slow down what health communication can be about. In the quote above, through her hesitation, Helena speculates rather than gives a clear-cut answer about how male and female subjects *in fact* care and get affected. This moment makes it possible to slow down care, including who can be cared for, and how people get affected. More specifically, Helena’s hesitation opens up an *ungendering* of cervical cancer, HPV vaccination and care.

Relational care instead of individual risk

The focus on care for relatives was contrasted with a focus on individual risk. Instead of addressing subjects by emphasizing risks linked to individual behaviors (as is common in health campaigns), it was emphasized that this campaign should depict care for others. In Klara’s words:

> Emotional messages about HPV vaccination could have been from a doctor who says that “if you don’t get vaccinated, you risk becoming one of the 80–100 who die every year”. That’s one way. But what happens then? Then we have actually lost the part that is a form of care and warmth toward each other. And that is what we implicitly express here […] We’re touching upon the emotional, that they actually can die. But it’s not merely to increase, what should I say, the risk

> level. You can communicate high risk level [...] Inducing a bit of fear. That you can do. But we chose to not do so as we wanted to actually keep the warm part and show that people take care of each other.

For Klara the difference between communicating a high risk level and the "I love me" campaign is care. Care is stated to be in contrast with a focus on the individual subject's statistical probability of developing cervical cancer. In contrast to other campaigns that are inducing fear by focusing on a "high risk level", Klara argued that the second "I love me" campaign evokes care and love. She positioned herself critically toward campaigns that use fear to moralize individual risk behaviors; campaigns telling people that *you* are at risk due to *your* behavior. In this vein, she emphasized that she was critical toward scaremongering. Communicating fear and risk, she emphasized, would not be a responsible communication strategy for a public agency. In Klara's reasoning, a care for others that is about warm feelings of love is present in the campaign, and this is what makes it not be about individual risk and fear – but a responsible endeavor. Thus, a distinction between the responsible communication of death and disease as *relational care* and an irresponsible focus on *individual risk* was put forward.

The emphasis on caring relationships has to do with an understanding that facts are not enough to convince the target group. It is too abstract for people to care about HPV vaccination if it is presented as science-as-epidemiology. Communicating only facts would mean *a loss of care*. "We had already told [the audience] how many die from cervical cancer in [the region]. That we have told people. That we knew. We had, you know, talked about that in campaigns. It was pointless", Klara argued. In a similar manner, Helena stated that they needed to be "more serious and emotional" as facts had not convinced the target group. That they had already told people how many died due to cervical cancer means that they had said it with epidemiology-derived numbers. Instead of telling people using numbers, they hoped that telling it through personal experiences would help. "Then it's the connection between, you know, one's own life that was also said [by the target group] could work", Klara said. The formulation *one's own life* means to address people as caring beings affected by their close ones' health, instead of as individuals "at risk" due to their own behavior, presented through abstract numbers.

Similar to Klara, Helena distanced the second "I love me" campaign from fear-based campaigns. She compared the second "I love me" campaign to campaigns in Australia that, according to her, were problematic in how they mainly served to scare people.

> But then there are these really awful examples from, for example, Australia [which is a country] that has developed many tobacco movies with rotten lungs and veins that are totally … these really gross things. And that we, you know, don't do in Sweden.

Helena stressed that the "I love me" campaign was emotionally strong in that it focused on how the whole family would be affected by a cervical cancer diagnosis. In contrast, and similar to Klara's emphasis on the problems with a focus on individual risk and fear, the campaigns in Australia were stressed to be about scaring people as they focused on how people put themselves at risk due to their own behavior. Picturing a distance to Australian campaigns (as irresponsible and problematic) enabled Helena to affirm the "I love me" campaign as a responsible form of health communication.

Thus, Helena emphasized Sweden as a country where fear-focused scaremongering campaigns are not possible. Similar to Klara's focus on the values and trademark of the county council as discussed earlier, the county council's identity as a Swedish public authority was stressed as a reason for why they did not use fear-based and individual risk messages. This taps into an imaginary of Sweden as a Nordic welfare state that takes care of its citizens. Sweden is articulated as a country of care. This is different from how others have emphasized a transformation in Sweden toward an increased self-responsibilization of care that includes a discourse of individualized risk and a moralization of health (see e.g. Dahl 2012; Törrönen and Tryggvesson 2015). For Helena, however, a welfare imaginary makes it possible to position the campaign as being about relational care rather than about individual fear.

Helena and Klara positioned themselves as critical toward health campaigns that were about individual risk behaviors. Both contrasted this with how the second "I love me" campaign is about a care for others and, therefore, relationships rather than individuals "at risk". This, for them, was a crucial matter of care that enabled them to work with the values of the county council, and therefore to be accountable and responsible toward the concerned public.

This form of accountability inhabits specificities regarding care. As I have shown, care was envisioned as about warm feelings of love and close relationships. As discussed in Chapters 7 and 8, this vision easily reproduces care as something innocent that is not about sometimes troubling politics. This was quite strikingly made clear when Klara stated that the council did not think about "where we are ideologically"; they think instead about what will be a fitting communication strategy for "the effect" they want to achieve and "the assignment" they have (which is to increase vaccination coverage for the age and gender group concerned). To assume that a message about care is not about ideology means that it is something imagined being outside of politics (or that its politics is not relevant). How care in the campaign, and their work with it, was figuring as a matter of political governance was made absent.

Still, I argue that it can be important to slow this down. To stay with Helena's and Klara's version of care, I argue, can be important in the sense of how it may, even if it has its problems, unsettle and complicate predominant assumptions about HPV vaccination campaigns as a matter of individualized, yet gendered, risk. Their definition of care as relational and non-gender specific enabled them to problematize what subjects are being cared for, and how these subjects can be cared for. It enabled them to address relationships instead of individuals "at risk" and to include a range of subjects as caring in the context of cervical cancer. Thus, by imagining it as a non-troubling matter of warm and loving feelings, care was allowed to complicate and unsettle what contemporary health campaigns may be about. This illustrates how it analytically can be important for researchers to point toward the risks of envisioning care as innocent and politically neutral, and to *simultaneously* highlight parts or moments that can unsettle predominant, and exclusionary, configurations.

An absent care for the herd

A care for the population or the herd was absent in how my interviewees talked about the campaign as being about a care for others. When I asked Helena about whether at the county council they talk about herd immunity and protecting the population in connection with the campaign, she answered:

> A little … But there I must say I'm not that knowledgeable. It feels a bit like a disease control question […] Yes, yeah, we talked a bit about it as we were trying to set the goal for this older group. As you know, we had [a goal of] 95 percent for [the age group born between] 93 [to] 98 but we said that up to 26, we should have 95 percent who know about the campaign. Not 95 percent that should get vaccinated […] [W]hat can you purely medically have as a level for it, you know, to make a difference in the future?

Interestingly, when I asked about the presence of the population, the focus of the conversation changed from being about people caring for each other to being about disease control, medical facts and statistics. The population became present in a specific version: as a matter of science-as-epidemiology. Any idea of people caring for the population or the herd was strikingly absent. That is, the population or the herd as something other than abstract and impersonal science-as-epidemiology was not made intelligible. Even if Helena would perhaps say that science-as-epidemiology is about care (I imagine she definitively would), it was not discussed as such. Thus, a boundary between science-as-epidemiology and care was implicitly drawn, where the former became abstract and distant and the latter something close and personal. This compares to the distinction made by the epidemiologist Emma (in Chapter 7) between girls caring for themselves and a focus on the population. In that chapter, a care for the herd was not articulated as a possible way to address girls as it was assumed to imply compulsion and control. Here, instead, a care for the herd was made unintelligible since the herd was being connected to abstract and distant statistics, and therefore distant from what care was envisioned to be about: warm feelings of love to close ones.

The absence of the population or the herd in the interviews can be related to, yet contrasted with, other studies emphasizing how HPV vaccination campaigns illuminate a "new politics of prevention" with its main focus on individual risk rather than population health and herd immunity (Wailoo et al. 2010). However, even if my interviewees did not highlight population health and herd immunity, the focus was not on individual risk. No dichotomy was articulated between the population and individual risk. Instead, caring relationships were environed as *another form* of vaccination message than population health *or* individual risk.

The absence of a care for the herd can fruitfully be contrasted with other recent Swedish vaccination campaigns. For example, information campaigns from the Swedish National Board of Health concerning the swine flu vaccination from 2010 were focused on herd immunity. "The vaccine Pandemrix that is given against the flu A(H1N1) gives a 90 percent protection against getting afflicted by flu. Thus, everyone who gets vaccinated does not get full protection; but if many get vaccinated we can still avoid the few who don't get full protection from getting the flu", reads one information pamphlet. In emphasizing that it is important to get vaccinated to make sure that others do not get the flu, herd immunity is implicitly referred to. So a care for the herd has recently been a communication strategy in Swedish vaccination campaigns. In contrast, my interviewees emphasized a care for *known* others as important in the second "I love me" campaign. It was not envisaged that people wanted to be vaccinated because they cared for enabling herd immunity or that they care for *unknown* others.

Conclusions: care as a solution to accountability trouble?

In this chapter, I have analyzed how my interviewees at Bredland County Council discussed the second "I love me" campaign. I have focused on how a campaign that was communicating death and disease was seen as a trouble, and have discussed how care in different ways was called upon as a solution to this trouble. Accountability *with* care was invoked as needed. This took different forms. A vision of authenticity as unmediated access to people's real stories and experiences was articulated as enabling death and disease to be a message about care. Stories told from authentic witnesses who really cared about the issue. Real stories, as *experiences from someone* and warm and loving feelings of care, were thus contrasted with abstract and distant science-as-epidemiology, and with cold money.

Demarcations for what care is, and can be, were drawn in the interviews. Care was discussed as being about attending to relationships to close ones (most often the family); something warm and loving; about authenticity and telling real stories (instead of, for example, using models); about the county council taking responsibility for and being accountable to the concerned public (by for example, separating commercial actors from those telling authentic stories); and about listening to the girls. These forms of care are about caring for others, and making others affected.

It was also emphasized what care is *not*: care was not about money, care was not about communicating individual risks and death rates without including relationships to close ones and, therefore, care was not about communicating fear without relationality and love. Hence, science-as-epidemiology was emphasized as *not* being about communicating care. Science-as-epidemiology was linked instead to matters of the population or the herd *and* to communicating death risks as individual risk and moralization of health. As part of that, the herd or the population was not being linked to care but was, on the contrary, implicitly *contrasted to* care. In this sense, care was contrasted to both a focus on the population and a focus on individual risk as both these possible modes of address were linked to abstract and "cold" science-as-epidemiology and not to "warm care". Instead close relationships were articulated as a "middle-level" between the individual and the population and where care can be expressed. This is in interesting contrast to the often prominent presence of individual risk in public discourses (and perhaps above all, campaign discourses) around HPV vaccination and in most public health campaigns both inside and outside this particular context. However, in putting forward this "middle-level", possible contingencies and specificities regarding HPV vaccination evidence are made absent, and care was left as something always loving, warm and desirable.

An important aim of this chapter has been to discuss care in terms of public accountability and responsibility. I have discussed how authenticity and care were invoked as making it possible to delimit the campaign from commercial actors, and to be accountable to the values of the county council. Care was in different ways invoked as the reason for why the campaign was a responsible endeavor, and thus in line with the values of the county council.

This chapter illuminates some of the navigations invoked to affirm the second "I love me" campaign as a caring endeavor. This puts attention on (specific versions of) care as a mode of governance that is drawn upon to affirm public accountability, and shows how accountability was articulated as being about caring for the concerned public. As others have pointed toward, this easily makes absent *how* one is asked to care and *what* one is asked to care about. That is, the politics of care is made absent as care is envisioned as something warm, loving and innocent that can be used as a solution and delineator for public accountability. However, by paying close attention to sometimes subtle and marginal moments that trouble the idea of care as a clear-cut solution for accountability troubles, I have shown how hesitations, tensions and absences generated navigations for when, how, and for whom, care can enable public accountability.

10. CONCLUSIONS:
Promising and Troubling Matters of Care

In this study I have focused on matters of care to enable an approach that "stays with" the complex promises and troubles of HPV vaccination campaigns. I have done so to challenge normative and exclusionary designs of health campaigns, and to enable, foster and strengthen alternatives. I have used a matters-of-care lens to foster and strengthen versions of care which can enable health communication to become a more caring and inclusive practice that allows space-time for differences and alternatives. In following the matters of care circulating in my empirical material, I have shown that care in a setting of health communication can be many things, and that different human and nonhuman actors may be involved in doing it. In this study care has ranged from girls caring for friends and communicators trying to learn about what girls care about, to digital and textual devices facilitating and troubling girls' and professionals' capacities to care, and visual re-presentations of care as love and happiness.

Based on my study's findings, I argue that my focus on care as a multidimensional, situated and more-than-human phenomenon is a promising approach to care in a context of health campaigns. My approach may allow for a more inclusive and nuanced approach to care in this setting compared to perspectives that limit their scope to how health campaigns encourage individuals to care for themselves and for other individuals, and which center on how the presence of care in health campaigns merely or primary serves as a governing strategy. Importantly, my study shows that health campaign practices also may include a range of other more caring matters of care, and that it may be important to try to further strengthen these.

In this chapter, I will focus on some of the many matters of care this study has included. In doing so, I will answer the first four research questions I introduced in Chapter 1: (1) How, and what, matters of care are articulated and mediated in the campaigns? (2) How, and what, matters of care are articulated by county council professionals working with the campaigns? (3) By attending to absent, marginal, neglected and alternative articulations and narratives, what other matters of care are made present? (4) By attending to different temporalities of care, how is it possible to trouble and disrupt normative and exclusionary links between care and time?

I will first discuss the predominant matters of care in the campaigns, and some of the absences these include. From there, I will move to the matters of care made present by my attention toward neglected and alternative articulations and narratives. I then turn to the predominant matters of care in my interviewees' narratives, followed by a discussion concerning marginal and alternative matters of care in the interviews. I will especially focus on temporalities of care. I have separated the campaigns from the interviews since the different empirical materials have generated partly different empirical findings. Whereas the campaign chapters have shown re-presentations and articulations of, for example, links between temporalities of care and different happy and unhappy feelings, the interview chapters have been centered on professionals' articulations concerning how to reach and listen to girls.

I then discuss the study's empirical contribution to research on health communication. The section ends with a few suggestions for how health campaigns can be designed in a more inclusive manner. The final section is focused on the study's contribution to STS theory concerning temporalities of care. This answers my study's final research question: (5) How can these findings contribute to STS discussions on matters of care in technoscience?

Gendered and future-oriented matters of care

A range of different matters of care are articulated and re-presented in the HPV app and the two "I love me" campaigns. The most prevalent ones, in one way or another, concern the *gendering* of care and the *temporalities* of care.

The gendering of care is done differently in the campaigns. As discussed in Chapter 4, the video in the HPV app encourages girl-centered self-care through often circulating gendered re-presentations of girls and women as feminine looking. Also, the presence of a caring female school nurse in the video reproduces a vision of gendered labors of care. Thus, as a device that is envisioned to facilitate girls' capacity to care for themselves, the HPV app enables care that involves gendered tropes. Similarly, and as was analyzed in Chapter 6, the first "I love me" campaign includes gendered images that promise happiness and love through a "pink-washing" of cervical cancer, girlhood and care. Here, cervical cancer as something often painful and scary is made absent in favor of a "rosy hopefulness" (Jain 2013: 86).

By linking self-care and friendship-care to these versions of love and happiness, girls are asked to care for the HPV vaccination, for themselves, and for others. This is re-presented in accordance with normative articulations of what it means to be a young women or girl. For example, the focus on a care for friends effectively makes absent sexual dimensions of HPV vaccination by re-presenting girls as caring for friends and not for presumptive sexual partners. This focus on love and happiness as matters of care equates care with positive, happy feelings about girl empowerment, and it articulates care as something innocent and always for the better.

Moving from the campaign images of the first "I love me" campaign to the conversations on the Facebook site, a gendering of care is still a widely circulating trope. For example, through the use of experiential knowledge of being a girl or a mother, different publics used the Facebook commenting and like devices to align themselves with the campaign message of "I love me", or to distance themselves from the same. Many mothers also argued for HPV vaccination as something positive on the basis of them caring for their daughter, something that reproduced a gendered trope of mothers' care responsibility for their children. Another example of how gendered tropes were enrolled was how ideas of women's empowerment were (in line with the empowerment vision of "I love me") used by young women or girls to affirm each other's vaccination decisions ("U goo girl!!!"), and to affirm getting vaccinated as a matter of girls caring for themselves.

Still related to a gendering of care, but re-presented in a different manner, the cancer narratives inhabit a wide register of feelings of care, which are also gendered. For example, the male subjects telling their stories often draw upon tropes about males being the strong ones who "hold it together". What is more, some of them tell stories about their fear of failure when not being able to stay strong any longer. That is, they tell stories about a care for their partners which is closely linked to a gendered idea of males as caring *through* protection, and by not showing their feelings.

In contrast to the focus on love and happiness in the campaign images of the first "I love me" campaign, the cancer narratives in the second campaign are linked to a wide register of unhappy feelings of care such as fear, pain and grief. For example, in the second "I love me" campaign, to care is articulated as a state of anxiety; as they care they are anxious about what will happen to their close ones.

Another gendered matter of care in the campaigns concerns articulations of sexual activity and sexual dimensions. Notably, in the HPV app, sexual dimensions of HPV vaccination are mainly made present through a focus on risk of sexual infection and the need to use protection. Therefore, and in contrast to other HPV vaccination studies, sexual dimensions are not made absent. Still, and as I argued in Chapter 4, how sexual dimensions are made present often includes gender and sexual politics. The HPV app asks girls to care for themselves through re-presentations of teenage sex as a matter of risk. Notably, this reproduces a "sex negativity" that makes present sex as a matter of risk, and makes absent how it can potentially be a source of curiosity, desire and pleasure for teenage girls.

In the HPV app many specificities regarding links between sexual activity, HPV and cervical cancer are made present. Yet in how facts are re-presented as neutral and stable, any ideas that the facts might change are made absent. In contrast to this, in the first "I love me" campaign's images, links between sexual activity, HPV and cervical cancer are made absent or marginalized (and as was later critiqued by the professionals at Bredland County Council themselves, and by girls). Making the sexual dimension of HPV absent serves to reproduce a gendered trope assuming that girls need to be protected from sexual matters. In the case of the cancer narratives, specificities and uncertainties regarding HPV vaccination and its links to sexual activity and cancer are made absent in favor of gendered narratives about people caring for each other.

The two "I love me" campaigns also re-present and articulate temporalities of care. The campaign storytellers in the second "I love me" campaign tell us that our close one might get afflicted by cancer in the future, and therefore we need to anticipate cervical cancer *now* to enable happy and healthy futures. This focus on that action is needed now includes articulations of care for others as an affective response and urgency; go home, spur your partner to get vaccinated now! Relatedly, in the first "I love me" campaign, many of the campaign images are oriented toward normative girlhoods and gendered happy futures through articulations stating that girls ought to care *now* for themselves and for others. This articulates care as a matter of immediacy; "get vaccinated now!" Thus, in the two "I love me" campaigns gendered matters of care are related to a specific temporality of care: care as an anticipatory immediacy.

Engaging neglected and alternative matters of care

Many of the matters of care figuring in the three campaigns reproduce normative articulations of, and links between, gender, sexuality, feelings and temporalities. Therefore, I have emphasized that it is crucial to ask "for whom?" the campaigns articulate and re-present matters of care, and to attend to the exclusions these matters of care enact. In trying to disrupt such articulations of care, I have pointed to many other, often marginal, matters of care which are also present, or that can be made present. Through a commitment toward marginal, neglected, absent and alternative articulations, several other matters of care have been explored. My aim has been to try to foster and strengthen alternatives to often normative and exclusionary matters of care, and by doing so disrupt and unsettle the tighter knots that hold dominant articulations in place.

In Chapter 4, I framed the discussion of the HPV app through a "care for neglected things" (Puig de la Bellacasa 2011). In doing so, I questioned why it seems easy to neglect seemingly simple apps like the HPV app, which provide information, and that do not generate data (as self-tracking apps do). Guided by this concern, I explored what temporal "alter-narrative" a care for the HPV app can enable.

With this approach, I emphasized that the HPV app inhabits temporal consistencies between "the new" and "the old". The app, I argued, is assembled through a range of coexisting "older" and "newer" media devices such as a video, a calendar, information pamphlets and a push notification service. Therefore, the HPV app does not only reproduce gendered tropes of "sex negativity" and girls and women as feminine looking, it *also* inhabits "temporal mess"; multilayered and coexisting temporalities of mediation. Based on this, I argue that a care for the HPV app provides an alter-narrative that troubles simplified techno-scientific timelines which privilege "the new" and that often are highly future-oriented.

During its existence, the Facebook "I love me" site inhabited temporalities of care that complicate a vision of HPV vaccination as something to do immediately. Whereas Bredland County Council and a vaccination-positive public often, and in line with the campaign images, articulated care as a matter of urgency and immediacy ("get vaccinated now!"), a vaccination-critical public encouraged people to slow down, think and learn more before making a vaccination decision. They encouraged people not to align themselves with the idea that girls needed to get vaccinated now, without having taken the time to think it through. In doing so, they also made present exclusionary dimensions of the "I love me" promise of care as love and happy futures, such as that this message came with a reduction of uncertainties. In this vein, through these critical comments, the Facebook site provoked alternative temporalities of care which troubled an idea that girls should get vaccinated now to safeguard a healthy and happy future. Based on this, I argued that a slower temporality of care facilitated a troubling of HPV vaccination as necessary and urgent, and opened up for uncertainties and thoughtfulness.

The Facebook site enabled diverse publics to engage in HPV vaccination matters. In doing so, it also allowed for alternative matters of care to trouble exclusionary articulations of care as love and happiness. Formulated in this way, the Facebook included "valuable moments during which different perspectives opened up" (Felt 2016: 193). Since the site allowed for alternatives to emerge, it also held a promise of "co-learning" between experts and publics in a setting of public engagement.

At the same time, a polarity between proponents and critics served at least partly to close down this possibility. That the focus often landed on moralizing assumptions about how people *should* care rather than how they *could* care, shows that the site was part of an already troubled world that included strong opinions and positionings regarding vaccination matters. Nevertheless, in trying to stay with the promise of co-learning, in this study I have argued that the trouble articulated on the Facebook site provides a glimpse of an alternative world. This allows for thinking (doing) co-learning in public participation as a modest endeavor that includes a slowing down of urgencies in order to allow space-time for curiosity, surprises and detours.

The Facebook site inhabited diverse articulations of happy and unhappy feelings, which also facilitated different temporalities of care. This serves as one example of how this study shows how HPV vaccination campaigns can re-present and provoke a wide register of feelings that are also linked to different temporalities. In contrast to the first "I love me" campaign's images, the Facebook site articulated a wide register of unhappy feelings, such as indignation, worry and fear. With the help of different Facebook devices, publics troubled the "I love me" message of care as love and happiness, and each other's statements, through articulations of unhappy feelings.

Following this, I have suggested that publics and devices on the Facebook site can be understood as care troublers. As such, they disrupt and unsettle dominant matters of care. In particular, the Facebook devices participated in troubling campaign images that "pink-washed" cervical cancer through a focus on future-oriented and gendered happiness and love. Therefore, an attention to the care collectives of devices and publics on the Facebook site enabled me to make present other matters of care than those including anticipatory immediacy and gendered visions of care as love and happiness.

Quite different from how alternative matters of care were already circulating among the actors of the "I love me" Facebook site, in my analysis of the second "I love me" campaign I staged "my care" by holding on to my own discomfort as inhabiting a potential for a more responsible retelling of campaign cancer narratives about relational care. In doing so, I worked with an approach where I tried to resist a "ready-made-explanation" for critiquing the affective message of the campaign as simply being about scaremongering. By retelling my own affective memories related to my fathers' cancer diagnosis, I highlighted how affective responses toward the campaign were potentially multiple.

In using my own cancer narrative to trouble the call for urgency articulated by the campaign narratives, and in allowing myself to be affected and moved by the cancer narratives in the campaign, I provided an alternative engagement. I used it to discuss what it might mean to respond carefully to affective cancer narratives as part of a campaign. This was an entrance point for discussing the risks of equaling a care for the other with appropriation of the other's feelings, and for discussing affective ethico-politics of circulating cancer tropes. By paying close attention to the multilayered dimensions of feelings in the campaign, I showed that the campaign cancer stories provided illuminative narratives about matters of care as an ethico-politics of responding carefully to others and to matters in society.

By engaging and attending to neglected and alternative matters of care in the campaigns, I have made present and facilitated disruptions of (visions of) gendered, anticipatory and immediate timelines. In doing so, I have tried to respond carefully and responsibly to others, and to the societal and political happenings that the campaigns inhabit and articulate.

Reaching and listening to the girls and views from somewhere as matters of care

Moving from the campaigns to the interviews, in these too a range of matters of care were articulated. One of the matters of care occurring more frequently was the importance of reaching the girls where they are. Notably, the HPV app, Facebook and the "I love me" trailer were articulated as care enablers that could reach girls where they are with the information they needed to be able to care for themselves through HPV vaccination. These devices were invoked as inhabiting a promise of increased HPV vaccination coverage (that is, a speeding up or intensification of vaccination). However, how they figured as care enablers differed in connection to the different devices.

The HPV app was articulated as a transmitter of information that girls needed to make a vaccination decision. The information needed was often articulated as objective and impartial science-as-epidemiology and was contrasted to girls' fears of side effects, and with myths and rumors assumed to cause that fear. The idea that the app could reach girls where they are with the information they needed is in line with a current trend in the area of health communication that embraces the participatory potential of digital devices to reach teenagers (see Keelan et al. 2010; Zimet et al. 2013).

The "I love me" Facebook site and the vaccination trailer were envisioned as arenas that would enable girls to ask questions, and the county council or other girls to answer these. These could, for example, be questions about possible side effects, but also about practical information. This was envisioned to help increase the vaccination coverage. Thus, these devices were articulated by my interviewees at Bredland County Council as devices for girl participation that could reach girls where they are, and enable the care they needed (HPV vaccination information and/or HPV vaccination shots).

As devices for girl participation, the capacity of the HPV app, the Facebook site and the vaccination trailer to enable county councils to listen to the girls was frequently articulated as another related matter of care. I have, for example, argued that caring for the girls in the context of the "I love me" vaccination trailer partly included a goal-oriented temporality of care. That is, listening to the girls in this context was often invoked to enable increased vaccination coverage. As such it included a vision of a speeding up of vaccination. Similar to many of the campaign images that I discussed earlier in this chapter, this links HPV vaccination to a trope of anticipatory immediacy.

The methodological devices discussed in this study were also envisioned to enable the county councils to listen to girls. The focus groups in the case of the HPV app and the evaluations in connection to the "I love me" campaigns were assumed by my interviewees to make it possible for the county councils to work with health campaign devices in line with what concerned girls and young women themselves cared about.

These devices (the methodological and the participatory) were articulated as matters of care in how they were envisioned to enable the county council to listen to and learn from concerned girls. This is a choreographed form of listening to girls that easily elicits what is perceived as relevant opinions for increasing the HPV vaccination coverage. By imagining that these devices can capture public opinion, specific publics are in fact made. As this study shows, this envisions learning from girls as a progressive activity involving a clear goal of public opinion as a stable referent that will ensure increased vaccination coverage (that is, intensified vaccination). Therefore, it also includes a vision of a progressive temporality of care.

Listening to and reaching the girls and young women as matters of care were interlinked to another matter of care: the importance for county councils as public authorities to take responsibility for the message they communicate to the public concerned. It was emphasized by many of my interviewees that they needed to listen to what the public needed and wanted. They needed to communicate values of care, and they needed to show that they care. In connection to the second "I love me" campaign these matters were particularly present. It was emphasized that to use a mode of address that included disease and death required a responsible communication strategy, and public accountability. Here, my interviewees articulated that listening to the girls and young women enabled accountable and responsible communication. A demarcation was made between accountable and responsible communication that had to do with caring for the public concerned, or, notably, communication that was assumed to scare people, and to blame the public for their actions. Also, since the county council had listened to the girls and young women concerning what they thought could work, they stressed that the campaign was about care.

I argue that how care was drawn upon in the interviews about the second "I love me" campaign relies on a vision of care as something warm and innocent that can be used as a clear-cut solution for affirming public accountability. Yet I also think it is important to emphasize that this focus on care as public accountability distributes responsibility to the county council, and not, as is commonly done in the context of HPV vaccination campaigns, to the girls. In a situation where individual responsibility for care often is the dominant articulation, how the county council's responsibility here is emphasized is important.

Another frequently occurring matter of care in the interviews concerned the information the campaigns were envisioned to communicate. Here, very different ideas were emphasized in Bredland and Mittland. My interviewees in Mittland often articulated neutral and factual information (science-as-epidemiology) as a clear-cut solution to counteract fears, myths and rumors. Different from this, in Bredland facts were often downplayed in favor of a focus on what girls and young women themselves cared about (once again, listening to the girls was a vital matter of care).

I have emphasized that even when information was imagined by the county council professionals as neutral and factual, it was not always invoked as a scientific view from nowhere. Notably, and in the context of the second "I love me" campaign, science-as-epidemiology was articulated as problematic in how it included a message that was too abstract and distant and not about care. Instead, cancer narratives as *experiences from someone* were invoked as an important matter of care, since these allowed for communication that re-presented how people care for each other, and also showed that the county council cared for the public concerned. Relatedly, in the context of the HPV app, the school nurse was articulated as communicating information through a caring *view from someone*.

Marginal disruptions and trouble within professionals' narratives

Similar to my focus on alternative and neglected matters of care in the campaigns, I have attended to alternative and marginal matters of care in the interviews. In doing so, I have explored what it might mean to engage a mode of analysis that slows down the plot to allow space-time for often marginal moments of friction, subtleties, hesitations and uncertainties in county council professionals' narratives. Therefore, focus has been on how disruptions invoked *from within* troubled the often otherwise coherent narratives. In particular, this study shows that a vision of the HPV app and the cancer narratives as solutions to increase the vaccination coverage was troubled by often marginal and subtle moments in the interviews.

Disruptive laughter, hesitations and precautionary wordings made present uncertainties and differences. In the case of the HPV app such articulations troubled the HPV app and science-as-epidemiology as clear-cut solutions to vaccine fears. This invoked the HPV app as a potentially failing matter of care since it perhaps was not "fun enough", and it made present HPV vaccination uncertainties (such as that facts are changing *uncertain certainties* and that there are *in essence* no side effects). Through marginal wordings, HPV vaccination communication work and science-as-epidemiology were invoked as matters of living with uncertainty rather than as about clear-cut solutions and given certainties.

Another example of disruptive hesitations that opened up for differences was how my interviewee Helena first articulated that a care for others in the context of HPV vaccination "perhaps is something feminine", but then was silent, hesitated and said that she does not know. This opened up for an *ungendering of* cervical cancer HPV vaccination and care, and it articulated an indeterminacy regarding who is to be cared for in the context of HPV vaccination.

Being marginal articulations in the interviews, such moments and subtleties can easily be ignored, overlooked, and trivialized. Instead, by allowing them space and time, I have shown that these moments of friction troubled narratives of science-as-epidemiology as a clear-cut solution, and unsettled articulations of gendered HPV vaccination politics. This study highlights that analytically paying attention to such subtle moments can make present uncertainties and indeterminacies in the midst of an articulated need to stress and confirm certainty.

Another important finding concerns the presence of devices as care troublers in interviews. Notably, the methodological devices of county council conducted campaign evaluations and an interview study aiming for increasing vaccination coverage did not merely elicit public opinion, they also figured in the interview narratives as troublemakers. In particular, they were articulated as helping trouble the design of the first "I love me" campaign. They allowed several of my interviewees from Bredland County Council to argue that how they cared in the first "I love me" campaign was problematic since it was not in line with what girls themselves cared about. More concretely, the devices were drawn upon to trouble the absence of sexual dimensions in the first "I love me" campaign, and thereby the county council's previous assumption that girls needed to be protected from sexual matters in a context of HPV vaccination. As discussed in Chapter 7, these devices therefore enabled Bredland County Council to problematize (without them perhaps knowing that they did so) a dominant narrative circulating widely in society that assumes that girls need to be protected from "sex negativity" (e.g. risks). This study thus shows that these methodological devices not only figured in the interviews as instrumental devices for eliciting of "representationalist" assumptions of girls "out there". The devices also facilitated a matter of care other than the dominant ones about care as protection from "sex negativity".

Relatedly, the interviewees stressed unanticipated moments in connection with the "I love me" campaign vaccination trailer. These had to do with being there with the girls and attending to the needs invoked in that very situation, and to learn from that situation. Therefore, the trailer figured as a pacing care device that allowed space-time for a slowing down of vaccination where girls' concerns, fears and anxieties were cared for. Instead of caring for intensifying HPV vaccination, the focus was on attending to what the girls needed in that very situation. As argued in Chapter 7, this slowing down allowed for a troubling of an urgency of vaccination, and inhabited space-time for other configurations of what public participation in health communication could be about.

Throughout, I have tried to take seriously what my interviewees care about and to follow what is enabled by *their* care. Thus, I have emphasized that even if it is important to problematize, for example, normative re-presentations of care as gendered care work and assumptions about care as something warm and always for the good, the same articulations can *also* do important work.

Bredland County Council's reformulation of what girls care about is one example of how my interviewees' own definition of care unsettled predominant articulations. As argued, the methodological devices of the interview evaluation and the interview study conducted by Emma enabled my interviewees to trouble a matter of care reproducing "sex negativity". Also, viewing care as something intrinsically good, made it possible for my interviewees (perhaps without knowing that they did so) to problematize predominant assumptions about HPV vaccination as a matter of individual risk, and also who the subjects who are being cared for in a context of HPV vaccination are. What is more, it was their definition of care that allowed the focus on public accountability to locate responsibility to Bredland County Council, and not primarily to girls and young women. Therefore, paying close attention to my interviewees' definitions of care has enabled me to highlight how a troubling of often circulating tropes was invoked *from within* their narratives, and, thus, from within their own practice.

Contribution to empirical field of study: care in health campaigns

Previous research about care in health campaigns emphasizes a care for the self and a (often gendered) care for others (see e.g. Lupton 1995; Serlin 2010b; Fraser and Seear 2011). My study contributes to this research by showing how also other matters of care than self-care and care for others may be involved in a context of health campaigns.

I will focus on three empirical contributions of this study. First, I discuss why it may be important to attend to what people working with health campaigns care about. Second, focus is put on some of the problems with reducing uncertainties and differences in campaign designs. Finally, I discuss public participation in health communication, and give some suggestions as to how health campaigns can be done differently.

County council professionals as people who care

The county council professions I have met and talked with work in the midst of a vaccination context with often polarized opinions about what is right and wrong. There are very different knowledge claims concerning vaccine certainties and uncertainties, and vaccine fears and hopes. It is a context that includes "intense feelings of concern" (Roberts 2015: 31). People working with health communication try to find ways of responding to people's fears and worries about vaccinations. If the current ways do not seem to work, they work hard to find better ways.

In a vaccination context of circulating uncertainties and fears, working out ways of allowing girls (and others) to participate in designing, discussing and responding to health communication can be important work. In line with Eidenskog's (2015: 198) argument, I therefore claim that approaching what my interviewees are doing through a lens of care "brings a sense of humbleness" for the challenges county council professionals are confronted with.

I have shed light on how experts working with public health communication themselves struggle with what might be good information to communicate to concerned publics. I have shown that complexities and contingencies may exist *within* practices of health communication, matters that are often made invisible when moved to the public arena of on-going campaigns. Following this, I argue that trying to better understand the nuances and complexities in health campaign practice is crucial for social scientists and/or humanities scholars, if they want to provide insights that can help make health campaigns become more inclusive and caring. By attending carefully to the difficulties and challenges professionals working with health communication are confronted with, it is also possible to put forward *situated* critique that takes into account what the tensions, stakes and conflicts involved might be.

I have also critiqued some of the reductions of complexities and uncertainties in the actual campaigns. I will now turn to this.

Lessons about uncertainties and exclusions

I have problematized re-presentations of gender and sexuality that reproduce exclusionary and normative articulations of how girls need to take care of themselves. For example, the focus on care as a matter of girl empowerment in the first "I love me" campaign is closely linked to gendered tropes assuming that girls need to be taught to love themselves and that they need to take responsibility for their own health. This easily serves to moralize girls' health actions, and ignore how these are conditioned on social and political inequalities and circumstances. As others have shown, it is possible to design campaigns that take into consideration how people's actions are embedded in social, political and material circumstances, and that aim for collective collaboration rather than individualized responsibility (see e.g. Dutta 2007; Vardeman-Winter 2014).

In the move from county council work to the public arena, nuances and complexities often disappear. This study shows that this can have problematic implications. I have, for example, highlighted how it is problematic to present HPV vaccines as vaccines against cervical cancer as this reproduces a normative assumption that girls need to be protected from the sexual dimensions of HPV vaccination.

In the HPV app sexual matters are first and foremost re-presented as a matter of risks of infection. This easily serves to reproduce a "sex negativity" which in the context of teenage girls tends to make sexual matters present in terms of first and foremost risks. Sexually related public health communication can instead be used as an opportunity to open up dialogue about sexual matters in a broader manner. Especially, my study indicates that it is important to extend what sexual dimensions of health communication can be about from risks to also include sexual activities as a potential source of desires, pleasures and hopes.

As I have discussed, scientific findings in the campaigns are often presented as stable and neutral facts. My study shows that how professionals working with the campaigns talk in interviews about such findings is more complex matters, and that they have a sensibility toward facts as something changing and variable.

Ways to think about public participation in health communication

Through my analysis of the "I love me" trailer and the Facebook site, it is evident that participatory initiatives can open up other perspectives and identities than those usually involved in health communication practices. Notably, I have discussed how the trailer enabled moments where girls could articulate matters not directly related to increasing vaccination coverage, but which were important to them. Relatedly, the diverse participants at the "I love me" Facebook site opened up perspectives often not included when using other health communication media.

Based on this, and in line with what other STS researchers working on public participation in technoscience have shown (see e.g. Michael 2012; Felt 2016), I argue that the "detours" allowed for on the Facebook site and by the trailer can be understood as promising. In relation to the Facebook site, some of the uncertainties that publics stressed helped problematize the gendered and individually focused message of the first "I love me" campaign, and illuminated problems with describing HPV vaccines as vaccines against cervical cancer. The "I love me" Facebook site included problematic dimensions since it often involved moralizing sentiments (from critiques, among others), but to fully dismiss the claims of vaccination-critical publics in a context of vaccination communication can also be problematic. It can serve to reproduce "a highly polarized public discourse that is not conductive to the sort of careful deliberation desirable for addressing complex issues" (Martin 2015: 155).

My study suggests that it is problematic to try to use participatory technologies such as social media and smartphone apps to provide publics with factual information to "counter-act" affective and experiential ways of understanding and debating vaccination. Such a model is widely critiqued in STS for resting on an assumption of a "deficit" in publics' understanding of science (see e.g. Bucchi and Neresini 2008). This model often builds upon an idea that publics misunderstand scientific findings, and that more information will solve the problem (see e.g.

Leach and Fairhead 2007). In a context of vaccination, this model might easily reproduce a polarized debate between experts emphasizing factual information, and publics responding with feelings and experiential knowledge. Because of how the current vaccination debate is organized, factual claims will in such a situation be taken seriously, and experiential and affective knowledge will be dismissed, trivialized or responded to with factual claims.

How could then health campaigns be done in a more inclusive and caring manner?

Health campaigns done differently

It is important to design campaigns that include a diversity of identities and perspectives. There is not one right solution for how this can be done, and which can be used universally. Instead it requires attention to the specificities of the context and publics concerned. I argue that it would be fruitful to create working groups where responsible public authorities invite humanities and/or social science researchers (such as gender studies or child studies scholars) and concerned publics into the design process from early on.

There is potential in working with technologies that facilitate public participation in health communication (such as social media and on-location events where experts and publics meet). My study indicates that this allows for a diversity of perspectives and for learning between experts and publics. However, my research shows that this is often done in too instrumental a manner that does not embrace the full potential of public participation in health communication. It becomes a paradox if enabling public participation in this context first and foremost has the goal of promoting specific health interventions and increasing public compliance.

My study suggests that it would be valuable to create more open-ended forms of public participation in health communication, which do not only serve to promote different health interventions. Such events can enable processes of co-learning between publics and experts that allow for surprising and unanticipated lessons and insights. This would mean that it would be crucial to facilitate space and time for events and initiatives that allow for diverging and different knowledge claims and which do not primarily steer publics into what is perceived as *the* right action. If public authorities want to learn from publics it is important to allow for publics to voice perspectives, ideas and opinions that may challenge experts' own convictions and pre-understandings. More open-ended arenas for public participation in health communication allow for such an endeavor. I understand that this would require a change in how public authorities' health communication work is conducted, a change in national policies, and extended resources would have to be put into this. But my research suggests it would be valuable since it allows for insights and perspectives easily overlooked or neglected in more goal-oriented and narrow approaches to public participation in health communication.

Theoretical and methodological contribution: temporalities of care

This study contributes to discussions concerning temporalities of care, including how such forms of care relate to ethico-politics, feelings and materialities of care. I have approached temporalities of care as an empirical doing, and as an analytical and ethico-political standpoint. As already mentioned at the beginning of this chapter, this means that I have attended to two interlinked dimensions of temporalities of care: those circulating among the actors I have studied, and those I have tried to foster and strengthen in my research. In doing so, I have discussed dominant articulations of care as an urgency (anticipatory immediacy), and moments of slowed down and coexisting temporalities of care. Here, I have drawn upon feminist work emphasizing links between care and time (Haraway 2011; Puig de la Bellacasa 2015; Schrader 2015), including how a slowed down mode of attention can be a matter of trying to foster more caring relations (Jerak-Zuiderent 2013, 2014; Martin et al. 2015). I have further developed this discussion in several ways.

Devices as enabling and troubling temporalities of care

I have developed two concepts for discussing how devices articulate and mediate temporalities of care. Borrowing the term "pacing devices" from digital media scholar Esther Weltevrede and colleagues (2014: 135), I have discussed temporal dimensions of care devices. I have attended to how devices (such as digital devices) can be fruitfully studied through the concept of "pacing care devices". I argue that this notion helps to explore how care devices participate in, and mediate, different, and coexisting, temporalities of care. Following this, and as the vaccination trailer and the Facebook social buttons did in this study, I argue that devices can mediate and enable coexisting paces of care; they can speed up, twist, turn and slow down care, and assemble different temporalities of care.

I have drawn upon Eidenskog's (2015) notion of "care enabler" to discuss how devices (such as the HPV app and Facebook social buttons did in this study) distribute and facilitate others' capacity to care. Based on my empirical findings, I have also developed the "sibling" concept "care troubler".

This concept has helped me study how devices participate in disrupting gendered visions of care *as* love and happiness and of care as an anticipatory immediacy. It has also enabled me to attend to how devices help professionals trouble how they previously cared for girls. Based on these findings, I argue that the notion of care troublers can help understand how devices facilitate others' capacity to problematize and unsettle configurations of care. In doing so, they can provoke other ways of doing care. In this respect, the notion of care troublers highlights care devices as temporal; they set things in motion, and unsettle current configurations. As a concept that allows for staying with the trouble of care, it may enable a focus on how disruptions are sparked by devices *from within* otherwise dominant configurations of care.

Coexisting temporalities of care

The focus on temporal dimensions of care devices is related to another area to which this study contributes: different temporalities of care as coexisting.

I have shown how hesitations, laughter and precautionary wordings in interviews can provide insights in this area. Disruptive moments can make present uncertainties, doubts and fears of failure. In doing so, they can also trouble anticipatory, goal-oriented and progressive temporalities of care. Since such disruption is evoked *from within* the dominant configuration, it is a matter of coexisting temporalities of care.

Whereas this is reminiscent of Jerak-Zuiderent's (2014) study on how disruptive laughter in interviews may open up for difference, I have engaged a discussion of what this can say about coexisting temporalities of care. Differently from Jerak-Zuiderent, I have shown how subtle disruptive laughter can trouble predominant temporalities of care by making present an always, already present lingering uncertainty. Following this, I suggest that paying close attention to (subtle) disruptive moments in interviews can help to trouble a dominant timeline *from within.*

Schrader hints at hesitations as opening up for more livable temporalities of care. She argues for the importance of slowing down temporalities to allow "space-times for hesitations" (Schrader 2015: 684). I have further developed what this might mean. In a vein similar to my focus on laughter, I have shown how an analytical space-time for subtle hesitations can make present uncertainties, indeterminacies and fears of failures (such as an ungendering of care, facts as changing and technoscientific solutions as not working). Therefore, I suggest that a focus on hesitations (articulated for example in interviews) can help to question normative visions that often include a dominant progressive and goal-oriented timeline.

I have also focused on how precautionary wordings can allow a space-time for uncertainties, and how this can disrupt visions of clear-cut and goal-oriented technoscientific solutions. Attention to precautionary wordings adds an example of what it concretely might mean to do "slower" research with the aim of disrupting normative timelines of care.

I agree with Jerak-Zuiderent (2013, 2014) and Schrader (2015) that theoretically and methodologically it can be valuable to use a slower mode of attention. Based on my study, I argue that paying attention to hesitations, laughter and precautionary wordings can be one way of taking seriously how a slower mode of attention might help disrupt for example productivist, anticipatory and goal-oriented temporalities of care.

Entangled feelings and temporalities of care

My study provides insights concerning the temporal dimensions of feelings of care. In particular, I have engaged in a theoretical discussion concerning how unhappy feelings can be linked to temporalities of care. This connection between unhappy feelings and temporalities is emphasized by Ahmed (2010: 186) in her conceptualization of care as a state of temporal anxiety; "to be full of care, to be careful, is to take care of things by becoming anxious about their future". Whereas Murphy (2015) draws upon Ahmed (2010) to stress the need for feminist STS care research on unhappy feelings, I have further elaborated on what unhappy feelings of care say about temporalities of care in technoscience.

In line with Ahmed's (2010) emphasis on the political and exclusionary dimensions of happiness, I have problematized anticipatory articulations of love and happiness in how they reduce complexities and uncertainties, and reproduce a gendering of care. What is more, I have attended to how unhappy feelings of discomfort, fear, worry and indignation can open up for alternative, and more inclusive, temporalities of care. Here, I have focused on how, for example, publics articulate unhappy feelings and how in doing so they problematize a vision of anticipatory immediacy, and articulate a slower temporality of thoughtfulness. Another example is how I "stayed with" my own discomfort as an entrance point for an alternative engagement in connection with the second "I love me" campaign's cancer narratives.

In showing how unhappy feelings such as discomfort, fear, indignation and worry can disrupt dominant timelines, and open up alternative and more inclusive temporalities of care, this study has further developed the recent critique toward a feminist theory tendency to equate care with happy or positive feelings (see Murphy 2015).

Future research

Following my suggestion concerning working with the notion of devices as care troublers, I suggest that exploring care devices as troublemakers can help researchers attend to how devices not only enable care, but also participate in disrupting and critiquing care. I encourage others to examine in their research devices that trouble care. Perhaps especially in settings that easily reproduce moralized articulations for how others should act, a focus on care troublers can make present how care is done in exclusionary ways, and open up alternatives that hopefully will enable more inclusive and livable lives.

Based on this study's findings concerning temporalities of care, I would suggest further research that pays close attention to the situated temporal specificities in how matters of care are made in technoscience. When studying future-oriented and gendered areas such as HPV vaccination campaigns, a focus on how slower and coexisting temporalities of care can disrupt anticipatory, immediate, gendered and heteronormative timelines may be vital. In other practices a range of other temporalities of care are likely to be involved, with their own promises and troubles. There is much to explore, challenge and cherish.

References

Adams, S., & Niezen, M. (forthcoming). Digital 'Solutions' to Unhealthy Lifestyle 'Problems': The Construction of Social and Personal Risks in the Development of eCoaches. *Health, Risk & Society.*

Adams, V., Murphy, M., & Clarke, A. (2009). Anticipation: Technoscience, Life, Affect, Temporality. *Subjectivity*, 28(1), 246–265.

Ahmed, S. (2004). *The Cultural Politics of Emotion.* Edinburgh: Edinburgh University Press.

Ahmed, S. (2007). The Happiness Turn. *New Formations*, 63, 7–14.

Ahmed, S. (2010). *The Promise of Happiness.* Durham, NC and London: Duke University Press.

Alvesson, M. (2011). *Intervjuer: Genomförande, tolkning och reflexivitet.* Malmö: Liber.

Aronowitz, R. (2010). Gardasil: A Vaccine Against Cancer and a Drug to Reduce Risk. In: Wailoo, K., Livingston, J., Epstein, S., & Aronowitz, R. (eds.) *Three Shots at Prevention. The HPV Vaccine and the Politics of Medicine's Simple Solutions.* Baltimore, MD: The Johns Hopkins University Press.

Aspers, P. (2007). *Etnografiska metoder: Att förstå och förklara samtiden.* Malmö: Liber.

Atkinson-Graham, M., Kenney, M., Ladd, K., Murray, C. M., & Simmonds, E. A. J. (2015). Care in Context: Becoming an STS Researcher. *Social Studies of Science*, 45(5), 738–748.

Axelsson, P. (2004). *Höstens spöke: De svenska polioepidemiernas historia.* Umeå University. PhD thesis.

Bäcklund, M. (2015). *Livmoderhalscancer (cervixcancer).* Available at: http://www.netdoktorpro.se/gynekologi-obstetrik/medicinska-oversikter/Livmoderhalscancer-cervixcancer/. Accessed: April 3, 2016.

Barad, K. (2007). *Meeting the Universe Halfway: Quantum Physics and The Entanglement of Matter and Meaning.* Durham, NC: Duke University Press.

Bell, K. (2012). Remaking the Self: Trauma, Teachable Moments, and The Biopolitics of Cancer Survivorship. *Culture, Medicine, and Psychiatry*, 36(4), 584–600.

Berlant, L. (2010). Cruel Optimism. In: Gregg, M., & Seigworth, G. J. (eds.) *The Affect Theory Reader.* Durham, NC and London: Duke University Press.

Betsch, C., Brewer, N. T., Brocard, P., Davies, P., Gaissmaier, W., Haase, N., & Rossmann, C. (2012). Opportunities and Challenges of Web 2.0 for Vaccination Decisions. *Vaccine*, 30(25), 3727–3733.

Birkbak, A. (2013). *From Networked Publics to Issue Publics: Reconsidering the Public/Private Distinction in Web Science.* Paper presented at the 5th Annual ACM Web Science Conference, Paris, May 2–4.

Björklund Larsen, L. (2010). *Illegal Yet Licit. Justifying Informal Purchases of Work in Contemporary Sweden.* Stockholm University. PhD thesis.

Bodén, L. (2015). The Presence of School Absenteeism. Exploring Methodologies for Researching the Material-Discursive Practice of School Absence Registration. *Cultural Studies <=> Critical Methodologies*, 15(3), 192–202.

Boero, N. C. (2010). Bypassing Blame: Bariatric Surgery and the Case of Biomedical Failure. In: Clarke, A., Mamo, L., Fosket, J. R., Fishman, J. R., & Shim, J. K. (eds.) *Biomedicalization. Technoscience, Health and Illness in the U.S.* Durham, NC and London: Duke University Press.

Bolter, J. D., & Grusin, R. (1996). Remediation. *Configurations*, 4(3), 311–358.

Borchorst, A., & Siim, B. (2002). The Women-Friendly Welfare States Revisited. *NORA: Nordic Journal of Women's Studies*, 10(2), 90–98.

Bowlby, S. (2012). Recognising the Time—Space Dimensions of Care: Caringscapes and Carescapes. *Environment and Planning A*, 44(9), 2101–2118.

Bragesjö, F., & Hallberg, M. (2009). *I forskningens närhet: En studie av MPR-kontroversens bakgrund och förvecklingar*. Nora: Nya Doxa.

Braun, L., & Phoun, L. (2010). HPV Vaccination Campaigns: Masking Uncertainty, Erasing Complexity. In: Wailoo, K., Livingston, J., Epstein, S., & Aronowitz, R. (eds.) *Three Shots at Prevention. The HPV Vaccine and the Politics of Medicine's Simple Solutions*. Baltimore, MD: The Johns Hopkins University Press.

Bredström, A. (2008). *Safe Sex, Unsafe Identities: Intersections of 'Race', Gender and Sexuality in Swedish HIV/AIDS Policy*. Linköping University. PhD thesis.

Brownlie, J., & Howson, A. (2005). 'Leaps of Faith' and MMR: An Empirical Study of Trust. *Sociology*, 39(2), 221–239.

Bucchi, M., & Neresini, F. (2008). Science and Public Participation. In: Hackett, E. J., Amsterdamska, O., Lynch, M., & Wajcman, J. (eds.) *The Handbook of Science and Technology Studies*. Cambridge, MA: MIT Press.

Burns, K., & Davies, C. (2015). Constructions of Young Women's Health and Wellbeing in Neoliberal Times: A Case Study of the HPV Vaccination Program in Australia. In: Wright, K., & McLeod, J. (eds.) *Rethinking Youth Wellbeing: Critical Perspectives*. Singapore: Springer.

Cancerfonden (2016). *Vård i världsklass – men inte för alla*. Avaliable at: https://www.cancerfonden.se/nyheter/vard-i-varldsklass-men-inte-for-alla. Accessed: April 3, 2016.

Cartwright, L. (1998). Community and the Public Body in Breast Cancer Media Activism. *Cultural Studies*, 12(2), 117–138.

Cartwright, L. (2008). *Moral Spectatorship: Technologies of Voice and Affect in Postwar Representations of the Child*. Durham, NC and London: Duke University Press.

Cartwright, L. (2013). How to Have Social Media in an Invisible Pandemic. In: Gates, K. (ed.) *The International Encyclopedia of Media Studies. Volume VI: Media Studies Futures*. Chichester: Wiley-Blackwell.

Cartwright, L. (2014). Visual Science Studies: Always, Already Material. In: Carusi, A., Hoel, A. S., Webmoor, T., & Woolgar, S. (eds.) *Visualization in the Age of Computerization*. New York and London: Routledge.

Casper, M., & Carpenter, L. (2008). Sex, Drugs, and Politics: The HPV Vaccine for Cervical Cancer. *Sociology of Health & Illness*, 30(6), 886–899.

Casper, M., & Carpenter, L. (2009a). Global Intimacies: Innovating the HPV Vaccine for Women's Health. *Women's Studies Quarterly*, 37(1–2), 80–100.

Casper, M., & Carpenter, L. (2009b). A Tale of Two Technologies: HPV Vaccination, Male Circumcision and Sexual Health. *Gender & Society*, 23(6), 790–816.

Casper, M. J., & Clarke, A. E. (1998). Making the Pap Smear into the "Right Tool" for the Job. Cervical Cancer Screening in the USA, circa 1940–95. *Social Studies of Science*, 28(2), 255–290.
Charles, N. (2013). Mobilizing the Self-Governance of Pre-Damaged Bodies: Neoliberal Biological Citizenship and HPV Vaccination Promotion in Canada. *Citizenship Studies*, 17(6–7), 770–784.
Charles, N. (2014). Injecting and Rejecting, Framing and Failing: The HPV Vaccine and the Subjectification of Citizens' Identities. *Feminist Media Studies*, 14(6), 1071–1089.
Clarke, A. (2005). *Situational Analysis: Grounded Theory after the Postmodern Turn*. Thousand Oaks, CA: Sage.
Coleman, R. (2015). Calculating Obesity, Pre-emptive Power and the Politics of Futurity: The Case of *Change4Life*. In: Amoore, L., & Piotukh, V. (eds.) *Algorithmic Life: Calculative Devices in the Age of Big Data*. London and New York: Routledge.
Colgrove, J. (2006). *State of Immunity: The Politics of Vaccination in Twentieth-Century America*. Berkeley, CA: University of California Press.
Collins, H., & Pinch, T. (2005). *Dr. Golem: How to Think About Medicine*. Chicago, IL and London: University of Chicago Press.
Connell, E., & Hunt, A. (2010). The HPV Vaccination Campaign: A Project of Moral Regulation in an Era of Biopolitics. *Canadian Journal of Sociology/Cahiers Canadiens de Sociologie*, 35(1), 63–82.
Coopmans, C., Vertesi, J., Lynch, M. E., & Woolgar, S. (2014). Introduction: Representation in Scientific Practice Revisited. In: Coopmans, C., Vertesi, J., Lynch, M. E., & Woolgar, S. (eds.) *Representation in Scientific Practice Revisited*. Cambridge, MA: MIT Press.
Cooter, R., & Stein, C. (2007). Coming into Focus: Posters, Power, and Visual Culture in the History of Medicine. *Medizinhistorisches Journal*, 42(2), 180–209.
Crawshaw, P. (2012). Governing at a Distance: Social Marketing and the (Bio)Politics of Responsibility. *Social Science & Medicine*, 75(1), 200–207.
Dahl, H. M. (2012). Neo-Liberalism Meets the Nordic Welfare State – Gaps and Silences. *NORA: Nordic Journal of Feminist and Gender Research*, 20(4), 283–288.
Davies, C. & Burns, K. (2014). Mediating Healthy Female Citizenship in the HPV Vaccination Campaigns. *Feminist Media Studies*, 14(5), 711–726.
Davies, S. R., & Horst, M. (2015). Crafting the Group: Care in Research Management. *Social Studies of Science*, 45(3), 371–393.
Diprose, R. (2008). Biopolitical Technologies of Prevention. *Health Sociology Review*, 17(2), 141–150.
Dugdale, A. (1999). Materiality: Juggling Sameness and Difference. In: Law, J., & Hassard, J. (eds.) *Actor Network Theory and After*. Oxford: Blackwell.
Durbach, N. (2004). *Bodily Matters: The Anti-Vaccination Movement in England, 1853–1907*. Durham, NC and London: Duke University Press.
Dutta, M. J. (2007). Communicating about Culture and Health: Theorizing Culture-Centered and Cultural Sensitivity Approaches. *Communication Theory*, 17(3), 304–328.
Edelman, L. (2004). *No Future: Queer Theory and the Death Drive*. Durham, NC and London: Duke University Press.
Egan, R. D., & Hawkes, G. L. (2008). Endangered Girls and Incendiary Objects: Unpacking the Discourse on Sexualization. *Sexuality & Culture*, 12(4), 291–311.

Eidenskog, M. (2015). *Caring for Corporate Sustainability*. Linköping University. PhD thesis.

Ekström, A., Jülich, S., Lundgren, F., & Wisselgren, P. (eds.) (2011). *History of Participatory Media: Politics and Publics, 1750–2000*. New York and London: Routledge.

Elam, M., & Gunnarsson, A. (2012). The Advanced Liberal Logic of Nicotine Replacement and the Swedish Invention of Smoking as Addiction. In: Larsson, B., Letell, M., & Thörn, H. (eds.) *Transformations of the Swedish Welfare State: From Social Engineering to Governance?* New York: Palgrave Macmillan.

Epstein, S. (1996). *Impure Science: AIDS, Activism, and the Politics of Knowledge*. Berkeley, CA: University of California Press.

Epstein, S. (2010). The Great Undiscussable: Anal Cancer, HPV, and Gay Men's Health. In: Wailoo, K., Livingston, J., Epstein, S., & Aronowitz, R. (eds.) *Three Shots at Prevention. The HPV Vaccine and the Politics of Medicine's Simple Solutions*. Baltimore, MD: The Johns Hopkins University Press.

Epstein, S., & Huff, A. (2010). Sex, Science and the Politics of Biomedicine: Gardasil in Comparative Perspective. In: Wailoo, K., Livingston, J., Epstein, S., & Aronowitz, R. (eds.) *Three Shots at Prevention. The HPV Vaccine and the Politics of Medicine's Simple Solutions*. Baltimore, MD: The Johns Hopkins University Press.

Evans, W. D. (2008). Social Marketing Campaigns and Children's Media Use. *The Future of Children*, 18(1), 181–203.

Evers, C. W., Albury, K., Byron, P., & Crawford, K. (2013). Young People, Social Media, Social Network Sites and Sexual Health Communication in Australia: "This is Funny, You Should Watch It". *International Journal of Communication*, 7, 1–20.

Expressen (2011). Vaccinstrulet kan kosta flickornas liv. *Expressen*, December 7. Available at: http://www.expressen.se/nyheter/dokument/vaccinstrulet-kan-kosta-flickornas-liv/. Accessed: March 5, 2016.

Federico, A. (2016). *Engagements with Close Reading*. New York: Routledge.

Felder, K., & Oechsner, S. (2015). On the Intertwinements of Care and Temporalities: Shared Reflections on Some of the Conference Themes. *EASST Review*, 34(4). Available at: https://easst.net/article/on-the-intertwinements-of-care-and-temporalities-shared-reflections-on-some-of-the-conference-themes/. Accessed: January 5, 2016.

Felt, U. (2016). The Temporal Choreographies of Participation: Thinking Innovation and Society from a Time-Sensitive Perspective. In: Chilvers, J., & Kearnes, M. (eds.) *Remaking Participation: Science, Environment and Emergent Publics*. New York and London: Routledge.

Fine, M. D. (2007). *A Caring Society? Care and the Dilemmas of Human Service in the Twenty-First Century*. New York: Palgrave Macmillan.

Foucault, M. (1988). The Ethic of Care for the Self as a Practice of Freedom. In: Bernauer, J., & Rasmussen, D. (eds.) *The Final Foucault*. Cambridge, MA: MIT Press.

Fraser, S., & Seear, K. (2011). *Making Disease, Making Citizens: The Politics of Hepatitis C*. London: Ashgate.

Freeman, E. (2010). *Time Binds: Queer Temporalities, Queer Histories*. Durham, NC and London: Duke University Press.

Fullagar, S. (2002). Governing the Healthy Body: Discourses of Leisure and Lifestyle Within Australian Health Policy. *Health*, 6(1), 69–84.

Gagnon, M., Jean, J., & Holmes, D. (2010). Governing through (In)Security: A Critical Analysis of a Fear-Based Public Health Campaign. *Critical Public Health*, 20(2), 245–256.

Gallop, J. (2000). The Ethics of Reading: Close Encounters. *Journal of Curriculum Theorizing*, 16(3), 7–17.
Gerlitz, C., & Helmond, A. (2013). The Like Economy: Social Buttons and the Data-Intensive Web. *New Media & Society*, 15(8), 1348–1365.
Gerlitz, C., & Lury, C. (2014). Social Media and Self-Evaluating Assemblages: On Numbers, Orderings and Values. *Distinktion: Scandinavian Journal of Social Theory*, 15(2), 174–188.
Gilligan, C. (1977). In a Different Voice: Women's Conceptions of Self and Morality. *Harvard Educational Review*, 47(4), 481–517.
Gilligan, C. (1982). *In a Different Voice: Psychological Theory and Women's Development*. Cambridge, MA: Harvard University Press.
Giraud, E., & Hollin, G. J. S. (forthcoming). Care, Laboratory Beagles and Affective Utopia. *Theory, Culture and Society*.
Gunnarsson, K. (2015). *Med önskan om kontroll: Figurationer av hälsa i skolors hälsofrämjande arbete*. Stockholm University. PhD thesis.
Gustafsson Reinius, L., Habel, Y., & Jülich, S. (2013). Att tänka med bussar: Några hållplatser. In: Gustafsson Reinius, L., Habel, Y., & Jülich, S. (eds.) *Bussen är budskapet: Perspektiv på mobilitet, materialitet och modernitet*. Stockholm: Mediehistoriskt arkiv.
Habel, Y. (2011). Say Milk, Say Cheese! Inscribing Public Participation in the Photographic Achieves of the National Milk Propaganda. In: Ekström, A., Jülich, S., Lundgren, F., & Wisselgren, P. (eds.) *History of Participatory Media: Politics and Publics, 1750–2000*. New York and London: Routledge.
Habel, Y. (2013). Mjölkpropagandans buss erövrar Sverige. In: Gustafsson Reinius, L., Habel, Y., & Jülich, S. (eds.) *Bussen är budskapet: Perspektiv på mobilitet, materialitet och modernitet*. Stockholm: Mediehistoriskt arkiv.
Haraway, D. (1991). *Simians, Cyborgs, and Women: The Reinvention of Nature*. London: Free Association Books.
Haraway, D. (1992). The Promises of Monsters: A Regenerative Politics for Inappropriate/d Others. In: Grossberg, L., Nelson, C., & Treichler, P. (eds.) *Cultural Studies*. New York: Routledge.
Haraway, D. (1997). *Modest_Witness@Second_Millennium.FemaleMan©_Meets_OncoMouse: Feminism and Technoscience*. New York: Routledge.
Haraway, D. (2004a). Cyborgs, Coyotes, and Dogs: A Kinship of Feminist Figurations and There Are Always More Things Going on Than You Thought! Methodologies as Thinking Technologies. In: Haraway, D. (ed.) *The Haraway Reader*. New York and London: Routledge.
Haraway, D. (2004b). Introduction: A Kinship of Feminist Figurations. In: Haraway, D. (ed.) *The Haraway Reader*. New York and London: Routledge.
Haraway, D. (2004c). Morphing in the Order: Flexible Strategies, Feminist Science Studies, and Primate Revisions. In: Haraway, D. (ed.) *The Haraway Reader*. New York and London: Routledge.
Haraway, D. (2008). *When Species Meet*. Minneapolis, MN: University of Minnesota Press.
Haraway, D. (2010). When Species Meet: Staying with the Trouble. *Environment and Planning D, Society and Space*, 28(1), 53–55.
Haraway, D. (2011). Speculative Fabulations for Technoculture's Generations: Taking Care of Unexpected Country. *Australian Humanities Review*, 50.

Harbers, H. (2010). Animal Farm Love Stories. About Care and Economy. In: Mol, A., Moser, I., & Pols, J. (eds.) *Care in Practice: On Tinkering in Clinics, Homes and Farms*. Bielefeld: Transcript.

Hariman, R., & Lucaites, J. L. (2007). *No Caption Needed: Iconic Photographs, Public Culture, and Liberal Democracy*. Chicago, IL and London: University of Chicago Press.

Harper, D. M., & Paavonen, J. (2008). Age for HPV Vaccination. *Vaccine*, 26, A7–A11.

Hemmings, C. (2005). Invoking Affect: Cultural Theory and the Ontological Turn. *Cultural Studies*, 19(5), 548–567.

Hildesheim, A., & Herrero, R. (2007). Human Papillomavirus Vaccine Should Be Given Before Sexual Debut for Maximum Benefit. *Journal of Infectious Diseases*, 196(10), 1431–1432.

Hill, M. J., Granado, M., Peters, R., Markham, C., Ross, M., & Grimes, R. M. (2013). A Pilot Study on the Use of a Smartphone Application to Encourage Emergency Department Patients to Access Preventive Services: Human Papillomavirus Vaccine as an Example. *Emergency Medicine and Health Care*, 1(1), 4.

Hirschfeld, L. A. (2002). Why Don't Anthropologists Like Children? *American Anthropologist*, 104(2), 611–627.

Hobson-West, P. (2003). Understanding Vaccination Resistance: Moving Beyond Risk. *Health, Risk & Society*, 5(3), 273–283.

Hobson-West, P. (2007). 'Trusting Blindly can be the Biggest Risk of All': Organised Resistance to Childhood Vaccination in the UK. *Sociology of Health & Illness*, 29(2), 198–215.

Hughes, B., McKie, L., Hopkins, D., & Watson, N. (2005). Love's Labours Lost? Feminism, the Disabled People's Movement and an Ethic of Care. *Sociology*, 39(2), 259–275.

Hunt, D., & Koteyko, N. (2015). 'What Was Your Blood Sugar Reading This Morning?' Representing Diabetes Self-Management on Facebook. *Discourse & Society*, 26(4), 445–463.

Jain, S. L. (2013). *Malignant: How Cancer Becomes Us*. Berkeley, CA: University of California Press.

Jain, S. L., & Stacey, J. (2015). On Writing About Illness: A Dialogue with S. Lochlann Jain and Jackie Stacey on Cancer, STS, and Cultural Studies. *Catalyst: Feminism, Theory, Technoscience*, 1(1), 1–29.

Jerak-Zuiderent, S. (2012). Certain Uncertainties: Modes of Patient Safety in Healthcare. *Social Studies of Science*, 42(5), 732–752.

Jerak-Zuiderent, S. (2013). *Generative Accountability: Comparing with Care*. Erasmus University. PhD thesis.

Jerak-Zuiderent, S. (2014). Keeping Open by Re-Imagining Laughter and Fear. *The Sociological Review*, 63(4), 897–921.

Jerak-Zuiderent, S. (2015). Accountability from Somewhere and for Someone: Relating with Care. *Science as Culture*, 24(4), 412–435.

Jethani, S. (2014). Mediating the Body: Technology, Politics and Epistemologies of Self. *Communication, Politics & Culture*, 47(3), 34–43.

Johannisson, K. (1994). The People's Health: Public Health Policies in Sweden. *Clio Medica*, 26, 165–182.

Johansen, V. F., Andrews, T. M., Haukanes, H., & Lilleaas, U. B. (2013). Symbols and Meanings in Breast Cancer Awareness Campaigns. *NORA: Nordic Journal of Feminist and Gender Research*, 21(2), 140–155.

Johansson Krafve, L. (2015). *Valuation in Welfare Markets: The Rule Books, Whiteboards and Swivel Chairs of Care Choice Reform*. Linköping University. PhD thesis.
Johnson, S. A. (2014). 'Maternal Devices', Social Media and the Self-Management of Pregnancy, Mothering and Child Health. *Societies*, 4(2), 330–350.
Juanita Brown, K. (2014). Regarding the Pain of the Other: Photography, Famine and the Transference of Affect. In: Brown, E., & Phu, T. (eds.) *Feeling Photography*. Durham, NC and London: Duke University Press.
Keelan, J., Pavri, V., Balakrishnan, R., & Wilson, K. (2010). An Analysis of the Human Papilloma Virus Vaccine Debate on MySpace Blogs. *Vaccine*, 28(6), 1535–1540.
Kember, S., & Zylinska, J. (2012). *Life after New Media: Mediation as a Vital Process*. Cambridge, MA: MIT Press.
Kenney, M. (2015). Counting, Accounting, and Accountability: Helen Verran's Relational Empiricism. *Social Studies of Science*, 45(5), 749–771.
Kittay, E. F. (1999). *Love's Labor: Essays on Women, Equality and Dependency*. New York and London: Routledge.
Knutson, T., & Öster, U. (2013). Utdragen upphandling försenade vaccinationer. *Advokaten*, 79(7). Available at: https://www.advokatsamfundet.se/Advokaten/Tidningsnummer/2013/Nr-7-2013-Argang79/Utdragen-upphandling-forsenade-vaccinationer--/. Accessed: March 5, 2016.
Korda, H., & Itani, Z. (2013). Harnessing Social Media for Health Promotion and Behavior Change. *Health Promotion Practice*, 14(1), 15–23.
Korp, P. (2004). *Hälsopromotion*. Lund: Studentlitteratur.
Kvale, S., & Brinkmann, S. (1997). *Den kvalitativa intervjun*. Lund: Studentlitteratur.
Läkemedelsvärlden (2016). Utreder om HPV-vaccin ska ges till pojkar. *Läkemedelsvärlden*, March 29. Available at: http://www.lakemedelsvarlden.se/nyheter/utreder-om-hpv-vaccin-ska-ges-till-pojkar-15064. Accessed: April 3, 2016.
Lappé, M. D. (2014). Taking Care: Anticipation, Extraction and the Politics of Temporality in Autism Science. *BioSocieties*, 9(3), 304–328.
Latimer, J., & Puig de la Bellacasa, M. (2013). Rethinking the Ethical in Bioscience: Everyday Shifts of Care in Biogerontology. In: Wrigley, A., & Priaulx, N. (eds.) *Ethics, Law and Society*. London: Ashgate.
Latimer, J., & Skeggs, B. (2011). The Politics of Imagination: Keeping Open and Critical. *The Sociological Review*, 59(3), 393–410.
Latour, B. (2004). Why Has Critique Run Out of Steam? From Matters of Fact to Matters of Concern. *Critical Inquiry*, 30(2), 225–248.
Latour, B. (2005). From Realpolitik to Dingpolitik. In: Latour, B., & Weibel, P. (eds.) *Making Things Public*. Cambridge, MA: MIT Press.
Law, J. (2009). Actor Network Theory and Material Semiotics. In: Turner, B. S. (ed.) *The New Blackwell Companion to Social Theory*. Chichester: Wiley-Blackwell.
Law, J., & Ruppert, E. (2013). The Social Life of Methods: Devices. *Journal of Cultural Economy*, 6(3), 229–240.
Law, J., & Singleton, V. (2005). Object Lessons. *Organization*, 12(3), 331–355.
Leach, M., & Fairhead, J. (2007). *Vaccine Anxieties: Global Science, Child Health and Society*. London: Earthscan.
Leask, J. (2002). *Understanding Immunisation Controversies*. University of Sidney. PhD thesis.
Leem, S. Y. (2016). The Anxious Production of Beauty: Unruly Bodies, Surgical Anxiety and Invisible Care. *Social Studies of Science*, 46(1), 34–55.

Lefebvre, C. (2009). Integrating Cell Phones and Mobile Technologies into Public Health Practice: A Social Marketing Perspective. *Health Promotion Practice*, 10(4), 490–494.

Leppo, A., Hecksher, D., & Tryggvesson, K. (2014). 'Why Take Chances?' Advice on Alcohol Intake to Pregnant and Non-Pregnant Women in Four Nordic Countries. *Health, Risk & Society*, 16(6), 512–529.

Levine, D. (2011). Using Technology, New Media, and Mobile for Sexual and Reproductive Health. *Sexuality Research and Social Policy*, 8(1), 18–26.

Lezaun, J., & Soneryd, L. (2007). Consulting Citizens: Technologies of Elicitation and the Mobility of Publics. *Public Understanding of Science*, 16(3), 279–297.

Liljeström, M., & Paasonen, S. (2010). Introduction: Feeling Differences – Affect and Feminist Reading. In: Liljeström, M., & Paasonen, S. (eds.) *Working with Affect in Feminist Readings: Disturbing Differences*. London: Routledge.

Lindén, L. (2013a). *Calculating Medical Uncertainties*. Tema T Working Paper, 350.

Lindén, L. (2013b). 'What do Eva and Anna Have to Do with Cervical Cancer?' Constructing Adolescent Girl Subjectivities in Swedish Gardasil Advertisements. *Girlhood Studies*, 6(2), 83–100.

Lindén, L., & Busse, S. (forthcoming). Two Shots at Prevention. Introducing HPV Vaccination in Austria. In: Johnson, E. (ed.) *Gendering Drugs: Feminist Studies of Pharmaceuticals*. London: Palgrave Macmillan.

Lindén, L., & Sullivan, M. (2015). Review of *Malignant: How Cancer Becomes Us* (University of California Press, 2013) and *Teratologies: A Cultural Study of Cancer* (Routledge, 1997). *Catalyst: Feminism, Theory, Technoscience*, 1(1), 1–21.

Lindström, K., & Ståhl, Å. (2014). *Patchworking Publics-in-the-Making: Design, Media and Public Engagement*. Malmö University. PhD thesis.

Löwy, I. (2010). HPV Vaccination in Context. A View from France. In: Wailoo, K., Livingston, J., Epstein, S., & Aronowitz, R. (eds.) *Three Shots at Prevention: The HPV Vaccine and the Politics of Medicine's Simple Solutions*. Baltimore, MD: The Johns Hopkins University Press.

Löwy, I. (2011). *A Woman's Disease: The History of Cervical Cancer*. Oxford: Oxford University Press.

Lukic, J., & Espinosa, A. S. (2011). Feminist Perspectives on Close Reading. In: Buikema, R., Griffin, G., & Lykke, N. (eds.) *Theories and Methodologies in Postgraduate Feminist Research: Researching Differently*. London: Routledge.

Lupton, D. (1995). *The Imperative of Health: Public Health and the Regulated Body*. London: Sage Publications.

Lupton, D. (1999). *Risk: New Directions and Perspectives*. Cambridge: Cambridge University Press.

Lupton, D. (2013). *Digitized Health Promotion: Personal Responsibility for Health in the Web 2.0 Era*. Working Paper No. 5. Sydney, NSW: Sydney Health & Society Group.

Lupton, D. (2014). Apps as Artefacts: Towards a Critical Perspective on Mobile Health and Medical Apps. *Societies*, 4(4), 606–622.

Lupton, D., & Jutel, A. (2015). 'It's Like Having a Physician in Your Pocket!' A Critical Analysis of Self-Diagnosis Smartphone Apps. *Social Science & Medicine*, 133, 128–135.

Lupton, D., & Thomas, G. M. (2015). Playing Pregnancy: The Ludification and Gamification of Expectant Motherhood in Smartphone Apps. *M/C Journal*, 18(5).

Lynch, M. (2014). Representation in Formation. In: Coopmans, C., Vertesi, J., Lynch, M. E., & Woolgar, S. (eds.) *Representation in Scientific Practice Revisited.* Cambridge, MA: MIT Press.
Maldonado Castañeda, O. J. (2015). *Making Evidence, Making Legitimacy: The Introduction of HPV (Human Papillomavirus) Vaccines in Colombia.* Lancaster University. PhD thesis.
Mamo, L., Nelson, A., & Clark, A. (2010). Producing and Protecting Risky Girlhoods. In: Wailoo, K., Livingston, J., Epstein, S., & Aronowitz, R. (eds.) *Three Shots at Prevention. The HPV Vaccine and the Politics of Medicine's Simple Solutions.* Baltimore, MD: The Johns Hopkins University Press.
Marks, N. J., & Russell, A. W. (2015). Public Engagement in Biosciences and Biotechnologies: Reflections on the Role of Sociology and STS. *Journal of Sociology*, 51(1), 97–115.
Marres, N. (2005). Issues Spark a Public into Being: A Key but Often Forgotten Point of the Lippmann-Dewey Debate. In: Latour, B., & Weibel, P. (eds.) *Making Things Public.* Cambridge, MA: MIT Press.
Marres, N. (2007). The Issues Deserve More Credit Pragmatist Contributions to the Study of Public Involvement in Controversy. *Social Studies of Science*, 37(5), 759–780.
Marres, N. (2015). *Brand or Issue? On Some Attempts to Analyse Digital Controversies Across Registers.* Paper presented at The Higher Seminar, Department of Thematic Studies, Technology and Social Change, Linköping University, October 14.
Martin, A., Myers, N., & Viseu, A. (2015). The Politics of Care in Technoscience. *Social Studies of Science*, 45(5), 625–641.
Martin, B. (2015). On the Suppression of Vaccination Dissent. *Science and Engineering Ethics*, 21(1), 143–157.
Maturo, A., & Setiffi, F. (forthcoming). The Gamification of Risk: How Health Apps Foster Self-Confidence and Why This is Not Enough. *Health, Risk & Society.*
McEwan, C., & Goodman, M. K. (2010). Place Geography and the Ethics of Care: Introductory Remarks on the Geographies of Ethics, Responsibility and Care. *Ethics, Place and Environment*, 13(2), 103–112.
McLuhan, M. (2009 [1964]). The Medium is the Message. In: Durham, M. G., & Kellner, D. M. (eds.) *Media and Cultural Studies: Keyworks.* Chichester: Wiley-Blackwell.
Michael, M. (2009). Publics Performing Publics: Of PiGs, PiPs and Politics. *Public Understanding of Science*, 18(5), 617–631.
Michael, M. (2012). "What Are We Busy Doing?" Engaging the Idiot. *Science, Technology & Human Values*, 37(5), 528–554.
Miele, M., & Evans, A. (2010). When Foods Become Animals: Ruminations on Ethics and Responsibility in Care-Full Practices of Consumption. *Ethics, Place and Environment*, 13(2), 171–190.
Milligan, C., & Wiles, J. (2010). Landscapes of Care. *Progress in Human Geography*, 34(6), 736–754.
Millington, B. (2014). Smartphone Apps and the Mobile Privatization of Health and Fitness. *Critical Studies in Media Communication*, 31(5), 479–493.
Mishra, A., & Graham, J. (2012). Risk, Choice, and the "Girl Vaccine": Unpacking Human Papillomavirus (HPV) Immunization. *Health, Risk & Society*, 14(1), 57–69.
Mol, A. (2002). *The Body Multiple: Ontology in Medical Practice.* Durham, NC and London: Duke University Press.

Mol, A. (2008). *The Logic of Care: Health and the Problem of Patient Choice.* London: Routledge.

Mol, A., Moser, I., & Pols, J. (eds.) (2010a). *Care in Practice: On Tinkering in Clinics, Homes and Farms.* Bielefeld: Transcript.

Mol, A., Moser, I. & Pols, J. (2010b). Care: Putting Practice into Theory. In: Mol, A., Moser, I., & Pols, J. (eds.) *Care in Practice: On Tinkering in Clinics, Homes and Farms.* Bielefeld: Transcript.

Moletsane, R., & Mitchell C. (2007). On Working with a Single Photograph. In: de Lange, N., Mitchell C., & Stuart, J. (eds.) *Putting People in the Picture: Visual Methodologies for Social Change.* Amsterdam: Sense.

Moulding, N. (2007). Love Your Body, Move Your Body, Feed Your Body: Discourses of Self-Care and Social Marketing in a Body Image Health Promotion Program. *Critical Public Health*, 17(1), 57–69.

Müller, R., & Kenney, M. (2014). Agential Conversations: Interviewing Postdoctoral Life Scientists and the Politics of Mundane Research Practices. *Science as Culture*, 23(4), 537–559.

Munro, E. (2013). "People Just Need to Feel Important, Like Someone is Listening": Recognising Museums' Community Engagement Programmes as Spaces of Care. *Geoforum*, 48, 54–62.

Murphy, M. (2015). Unsettling Care: Troubling Transnational Itineraries of Care in Feminist Health Practices. *Social Studies of Science*, 45(5), 717–737.

Myong, L., & Bissenbakker, M. (2016). Love without Borders? White Transraciality in Danish Migration Activism. *Cultural Studies*, 30(1), 129–146.

National Cancer Institute (2016). *HPV and Cancer.* Available at: http://www.cancer.gov/about-cancer/causes-prevention/risk/infectious-agents/hpv-fact-sheet. Accessed: March 5, 2016.

Neiger, B. L., Thackeray, R., Van Wagenen, S. A., Hanson, C. L., West, J. H., Barnes, M. D., & Fagen, M. C. (2012). Use of Social Media in Health Promotion Purposes. Key Performance Indicators, and Evaluation Metrics. *Health Promotion Practice*, 13(2), 159–164.

Newstead, C. (2009). Pedagogy, Post-Coloniality and Care-Full Encounters in the Classroom. *Geoforum*, 40(1), 80–90.

Nutbeam, D. (2000). Health Literacy as a Public Health Goal: A Challenge for Contemporary Health Education and Communication Strategies into the 21st Century. *Health Promotion International*, 15(3), 259–267.

Oinas, E., & Collander, A. (2007). Tjejgrupper: rosa rum, pippifeminism, hälsofrämjande? In: Oinas, E., & Ahlbeck-Rehn, J. (eds.) *Kvinnor, kropp och hälsa.* Lund: Studentlitteratur.

Olsson, U. (1997). *Folkhälsa som pedagogiskt projekt: Bilden av hälsoupplysning i statens offentliga utredningar.* Uppsala University. PhD thesis.

Ostherr, K. (2013). *Medical Visions: Producing the Patient through Film, Television, and Imaging Technologies.* Oxford: Oxford University Press.

Paasonen, S. (2007). Strange Bedfellows: Pornography, Affect and Feminist Reading. *Feminist Theory*, 8(1), 43–57.

Paasonen, S. (2010). Disturbing, Fleshy Texts: Close Looking at Pornograpghy. In: Liljeström, M., & Paasonen, S. (eds.) *Working with Affect in Feminist Readings: Disturbing Differences.* London: Routledge.

Palmblad, E., & Eriksson, B. E. (2014). *Kropp och politik: Hälsoupplysning som samhällsspegel.* 2nd ed. Stockholm: Carlsson.
Paul, K. T. (2016). "Saving Lives": Adapting and Adopting Human Papilloma Virus (HPV) Vaccination in Austria. *Social Science & Medicine*, 153, 193–200.
Pearce, L. (1997). *Feminism and the Politics of Reading.* London: Arnold.
Pedwell, C., & Whitehead, A. (2012). Affecting Feminism: Questions of Feeling in Feminist Theory. *Feminist Theory*, 13(2), 115–129.
Penkler, M. (2015). Quantifying the Body and Health: Adding to the Buzz? *EASST Review*, 34(4). Available at: https://easst.net/article/quantifying-the-body-and-health-adding-to-the-buzz-2/. Accessed: February 15, 2016.
Pérez-Bustos, T. (2014). Of Caring Practices in the Public Communication of Science: Seeing Through Trans Women Scientists' Experiences. *Signs*, 39(4), 857–866.
PHAS, Public Health Agency of Sweden (2014). *Statistik för HPV-vaccinationer.* Available at: http://www.folkhalsomyndigheten.se/documents/smittskydd-sjukdomar/vaccinationer/HPV_vaccination_tom_14-12-31.pdf. Accessed: March 5, 2016.
Pols, J. (2010). Caring Devices: About Warmth, Coldness and 'Fit'. *Medische Antropologie*, 22(1), 143–160.
Poltorak, M., Leach, M., Fairhead, J., & Cassell, J. (2005). "MMR Talk" and Vaccination Choices: An Ethnographic Study in Brighton. *Social Science & Medicine*, 61(3), 709–719.
Polzer, J., & Knabe, S. (2009). Good Girls Do … Get Vaccinated: HPV, Mass Marketing and Moral Dilemmas for Sexually Active Young Women. *Journal of Epidemiology & Community Health*, 63(11), 869–870.
Polzer, J. C., & Knabe, S. M. (2012). From Desire to Disease: Human Papillomavirus (HPV) and the Medicalization of Nascent Female Sexuality. *Journal of Sex Research*, 49(4), 344–352.
Porroche-Escudero, A. (2014). Perilous Equations? Empowerment and the Pedagogy of Fear in Breast Cancer Awareness Campaigns. *Women's Studies International Forum*, 47, 77–92.
Puig de la Bellacasa, M. (2010). Ethical Doings in Naturecultures. *Ethics, Place, and Environment*, 13(2), 151–169.
Puig de la Bellacasa, M. (2011). Matters of Care in Technoscience: Assembling Neglected Things. *Social Studies of Science*, 41(1), 85–106.
Puig de la Bellacasa, M. (2012). 'Nothing Comes Without its World': Thinking with Care. *The Sociological Review*, 60(2), 197–216.
Puig de la Bellacasa, M. (2014). Encountering Bioinfrastructure: Ecological Struggles and the Sciences of Soil. *Social Epistemology*, 28(1), 26–40.
Puig de la Bellacasa, M. P. (2015). Making Time for Soil: Technoscientific Futurity and the Pace of Care. *Social Studies of Science*, 45(5), 691–716.
Raghuram, P., Madge, C., & Noxolo, P. (2009). Rethinking Responsibility and Care for a Postcolonial World. *Geoforum*, 40(1), 5–13.
Ralph, L. J., Berglas, N. F., Schwartz, S. L., & Brindis, C. D. (2011). Finding Teens in Their Space: Using Social Networking Sites to Connect Youth to Sexual Health Services. *Sexuality Research and Social Policy*, 8(1), 38–49.
Rappert, B. (2015). Sensing Absence: How to See What Isn't There in the Study of Science and Security. In: Rappert, B., & Balmer, B. (eds.) *Absence in Science, Security and Policy: From Research Agendas to Global Strategy.* New York: Palgrave Macmillan.

Rappert, B., & Bauchspies, W. K. (2014). Introducing Absence. *Social Epistemology*, 28(1), 1–3.
Rehnqvist, N, Rosén, M., & Vilhelmsdotter Allander, S. (2008). Upphaussat cancervaccin svek mot kvinnors hälsa. *Dagens Nyheter*, January 31. Available at: http://www.dn.se/debatt/upphaussat-cancervaccin-svek-mot-kvinnors-halsa/. Accessed: April 3, 2016.
Reardon, J., Metcalf, J., Kenney, M., & Barad, K. (2015). Science & Justice: The Trouble and the Promise. *Catalyst: Feminism, Theory, Technoscience*, 1(1), 1–48.
Reich J. (2014). Neoliberal Mothering and Vaccine Refusal: Imagined Gated Communities and the Privilege of Choice. *Gender & Society*, 28, 679–704.
Reid, R. (2005). *Globalizing Tobacco Control: Anti-Smoking Campaigns in California, France, and Japan*. Bloomington and Indianapolis, IN: Indiana University Press.
Renold, E., & Ringrose, J. (2011). Schizoid Subjectivities? Re-Theorizing Teen Girls' Sexual Cultures in an Era of "Sexualization". *Journal of Sociology*, 47(4), 389–409.
Renold, E., & Ringrose, J. (2013). Feminisms Re-Figuring "Sexualisation", Sexuality and "The Girl". *Feminist Theory*, 14(3), 247–254.
RFSU, Riksförbundet för Sexuell Upplysning (2011). *Skandal att unga tjejer får vänta på vaccin*. Available at: http://www.rfsu.se/sv/Om-RFSU/Press/Pressmeddelanden/2011/Skandal-att-unga-tjejer-far-vanta-pa-vaccin/. Accessed: March 5, 2016.
Rich, E., & Miah, A. (2014). Understanding Digital Health as Public Pedagogy: A Critical Framework. *Societies*, 4(2), 296–315.
Rivano Eckerdal, J. (2015). Förståelser av HPV-vaccin mellan hälsopanik och tillit. *Socialmedicinsk tidskrift*, 92(6), 736–748.
Roberts, C. (2015). *Puberty in Crisis: The Sociology of Early Sexual Development*. Cambridge: Cambridge University Press.
Rose, H. (1983). Hand, Brain, and Heart: A Feminist Epistemology for the Natural Sciences. *Signs*, 9(1), 73–90.
Rose, H. (1994). *Love, Power, and Knowledge: Towards a Feminist Transformation of the Sciences*. Cambridge: Polity Press.
Rousseau, S. (2015). Is Sharing Caring? Social Media and Discourse of Healthful Eating. In: Abbots, E. J., Lavis, A., & Attala, M. L. (eds.) *Careful Eating: Bodies, Food and Care*. Farnham, Surrey: Ashgate.
Ruppert, E., Law, J., & Savage, M. (2013). Reassembling Social Science Methods: The Challenge of Digital Devices. *Theory, Culture & Society*, 30(4), 22–46.
Ruppert, E., Harvey, P., Lury, C., Mackenzie, A., McNally, R., Baker, S. A., & Lewis, C. (2015). *Background: A Social Framework for Big Data*. CRESC Working Paper Series, (138).
Sandin, B., & Halldén, G. (eds.) (2003). *Barnets bästa: En antologi om barndomens innebörder och välfärdens organisering*. Höör: Brutus Östlings förlag Symposium.
Savage, M. (2013). The "Social Life of Methods": A Critical Introduction. *Theory, Culture & Society*, 30(4), 3–21.
Schrader, A. (2015). Abyssal Intimacies and Temporalities of Care: How (Not) to Care about Deformed Leaf Bugs in the Aftermath of Chernobyl. *Social Studies of Science*, 45(5), 665–690.
Serlin, D. (ed.) (2010a). *Imagining Illness: Public Health and Visual Culture*. Minneapolis, MN: University of Minnesota Press.
Serlin, D. (2010b). Introduction: Towards a Visual Culture of Public Health. In: Serlin, D. (ed.) *Imagining Illness: Public Health and Visual Culture*. Minneapolis, MN: University of Minnesota Press.

Sevenhuijsen, S. (1998). *Citizenship and the Ethics of Care: Feminist Considerations on Justice, Morality, and Politics*. London and New York: Routledge.
Sevenhuijsen, S. (2003). The Place of Care: The Relevance of the Feminist Ethic of Care for Social Policy. *Feminist Theory*, 4(2), 179–197.
Singleton, V. (1995). Networking Constructions of Gender and Constructing Gender Networks: Considering Definitions of Women in the British Cervical Screening Programme. In: Grint, K., & Gill, R. (eds.) *The Gender-Technology Relation: Contemporary Theory and Research*. Bristol: Taylor & Francis.
Singleton, V. (1998). Stabilizing Instabilities: The Role of the Laboratory in the United Kingdom Cervical Screening Programme. In: Berg, M., & Mol, A. (eds.) *Differences in Medicine: Unraveling Practices, Techniques, and Bodies*. Durham, NC and London: Duke University Press.
Singleton, V. (2010). Good Farming. Control or Care? In: Mol, A., Moser, I., & Pols, J. (eds.) *Care in Practice: On Tinkering in Clinics, Homes and Farms*. Bielefeld: Transcript.
Singleton, V. (2012). When Contexts Meet: Feminism and Accountability in UK Cattle Farming. *Science, Technology and Human Values*, 37(4), 404–443.
Singleton, V., & Law, J. (2013). Devices as Rituals: Notes on Enacting Resistance. *Journal of Cultural Economy*, 6(3), 259–277.
Singleton, V., & Michael, M. (1993). Actor-Networks and Ambivalence: General Practitioners in the UK Cervical Screening Programme. *Social Studies of Science*, 23(2), 227–264.
Skeggs, B. (1999). Seeing Differently: Ethnography and Explanatory Power. *The Australian Educational Researcher*, 26(1), 33–53.
SKL, Swedish Association of Local Authorities and Regions (2010a). *NLT gruppens rekommendation till landstingen gällande HPV vaccination*. Available at: http://www.janusinfo.se/Documents/Nationellt_inforande_av_nya_lakemedel/HPV-100312.pdf. Accessed: March 5, 2016.
SKL, Swedish Association of Local Authorities and Regions (2010b). *Rekommendationer till landstingen om HPV-vaccin*. Available at: http://epi.vgregion.se/upload/Läkemedel/HPV/SKL_REK.pdf. Accessed: March 5, 2016.
Söderberg, E. (2011). The Pippi-Attitude as a Critique of Norms and as a Means of Normalization: From Modernist Negativity to Neoliberal Individualism. In: Fahlgren, S., Johansson, A., & Mulinari, D. (eds.) *Normalization and Outsiderhood: Feminist Readings of a Neoliberal Welfare State*. Sharjah, United Arab Emirates: Bentham Science Publishers.
Sparrman, A. (2014). Access and Gatekeeping in Researching Children's Sexuality: Mess in Ethics and Methods. *Sexuality & Culture*, 18(2), 291–309.
Spratt, J., Shucksmith, J., Philip, K., & McNaughton, R. (2013). Active Agents of Health Promotion? The School's Role in Supporting the HPV Vaccination Programme. *Sex Education*, 13(1), 82–95.
Stacey, J. (1997). *Teratologies: A Cultural Study of Cancer*. New York: Routledge.
Stein, S. (1991). *The Rhetoric of the Colorful and the Colorless: American Photography and Material Culture between the Wars*. Yale University. PhD thesis.
Stöckl, A. (2010). Public Discourses and Policymaking: The HPV Vaccination from the European Perspective. In: Wailoo, K., Livingston, J., Epstein, S., & Aronowitz, R. (eds.) *Three Shots at Prevention. The HPV vaccine and the Politics of Medicine's Simple Solutions*. Baltimore, MD: The Johns Hopkins University Press.

Sturken, M., & Cartwright, L. (2009). *Practices of Looking: An Introduction to Visual Culture.* 2nd ed. New York: Oxford University Press.
Sundén, J. (2012). Desires at Play on Closeness and Epistemological Uncertainty. *Games and Culture*, 7(2), 164–184.
Swedish Dental and Pharmaceutical Benefits Agency (2016). *Så fungerar högkostnadsskyddet*. Available at: http://www.tlv.se/lakemedel/hogkostnadsskyddet/sa-fungerar-hogkostnadsskyddet/. Accessed: March 5, 2016.
Swedish Government (1982). *Hälso- och sjukvårdslag.* Available at: https://www.riksdagen.se/sv/Dokument-Lagar/Lagar/Svenskforfattningssamling/Halso--och-sjukvardslag-1982_sfs-1982-763/. Accessed: March 5, 2016.
Swedish Government (1996). Prop. 1996/1997:60. *Prioriteringar inom hälso- och sjukvården*. Available at: http://www.riksdagen.se/sv/Dokument-Lagar/Forslag/Propositioner-och-skrivelser/Prioriteringar-inom-halso--och_GK0360/?text=true. Accessed: March 5, 2016.
Swedish Government (2007). Prop. 2007/08:110. *En förnyad folkhälsopolitik.* Available at: http://www.regeringen.se/rattsdokument/proposition/2008/03/prop.-200708110/. Accessed: March 6, 2016.
Swedish Government (2008). Prop. 2008/09:74. *Vårdval i primärvården.* Available at: http://www.regeringen.se/rattsdokument/proposition/2008/12/prop.-20080974/. Accessed: March 6, 2016.
Tan, Q. H. (2015). Governing Futures and Saving Young Lives: Willful Smoking Temporalities and Subjectivities. *Play, Recreation, Health and Well Being*, 9, 1–13.
Taylor, J. (2010). Queer Temporalities and the Significance of "Music Scene" Participation in the Social Identities of Middle-Aged Queers. *Sociology*, 44(5), 893–907.
Thelen, T. (2015). Care as Social Organization: Creating, Maintaining and Dissolving Significant Relations. *Anthropological Theory*, 15(4), 497–515.
Thorsén, D. (2013). *Den svenska aidsepidemin: Ankomst, bemötande, innebörd.* Uppsala University. PhD thesis.
Till, C. (2014). Exercise as Labour: Quantified Self and the Transformation of Exercise into Labour. *Societies*, 4(3), 446–462.
Torell, U. (2002). *Den rökande människan: Bilden av tobaksbruk i Sverige mellan 1950-och 1990-tal.* Stockholm: Carlssons Bokförlag.
Törrönen, J., & Tryggvesson, K. (2015). Alcohol, Health, and Reproduction: An Analysis of Swedish Public Health Campaigns against Drinking during Pregnancy. *Critical Discourse Studies*, 12(1), 57–77.
Towghi, F. (2013). The Biopolitics of Reproductive Technologies beyond the Clinic: Localizing HPV Vaccines in India. *Medical Anthropology*, 32(4), 325–342.
Tronto, J. (1993). *Moral Boundaries: A Political Argument for an Ethic of Care.* New York and London: Routledge.
Tronto, J. (2003). Time's Place. *Feminist Theory*, 4(2), 119–138.
Tulloch, J., & Lupton, D. (1997). *Television, AIDS and Risk: A Cultural Studies Approach to Health Communication.* St. Leonards: Allen & Unwin.
US Department of Health (2016). *Community Immunity ("Herd Immunity").* Available at: http://www.vaccines.gov/basics/protection/. Accessed: March 5, 2016.
Van Dijck, J. (2008). The Computer as Memory Machine. In: Smelik, A., & Lykke, N. (eds.) *Bits of Life: Feminism at the Intersections of Media, Bioscience and Technology.* Seattle, WA: University of Washington Press.

Van Dijck, J. (2011). Facebook as a Tool for Producing Sociality and Connectivity. *Television & New Media*, 13(2), 160–176.
Van Dijck, J. (2013). *The Culture of Connectivity: A Critical History of Social Media*. Oxford: Oxford University Press.
Vardeman-Winter, J. (2012). Medicalization and Teen Girls' Bodies in the Gardasil Cervical Cancer Vaccine Campaign. *Feminist Media Studies*, 12(2), 281–304.
Vardeman-Winter, J. (2014). Issues of Representation, Reflexivity, and Research-Participant Relationships: Doing Feminist Cultural Studies to Improve Health Campaigns. *Public Relations Inquiry*, 3(1), 91–111.
Verran, H. (2010). Number as an Inventive Frontier in Knowing and Working Australia's Water Resources. *Anthropological Theory*, 10(1–2), 171–178.
Verran, H. (2011). Number as Generative Device: Ordering and Valuing Our Relations with Nature. In: Lury, C., & Wakeford, N. (eds.) *Inventive Methods: The Happening of the Social*. London: Routledge.
Viseu, A. (2015). Caring for Nanotechnology? Being an Integrated Social Scientist. *Social Studies of Science*, 45(5), 642–664.
Wagner, L. C. (2005). "It's for a Good Cause": The Semiotics of the Pink Ribbon for Breast Cancer in Print Advertisements. *Intercultural Communication Studies*, 14(3), 210.
Wailoo, K., Livingston, J., Epstein, S., & Aronowitz, R. (2010). *Three Shots at Prevention: The HPV Vaccine and the Politics of Medicine's Simple Solutions*. Baltimore, MD: The Johns Hopkins University Press.
Weltevrede, E., Helmond, A., & Gerlitz, C. (2014). The Politics of Real-Time: A Device Perspective on Social Media Platforms and Search Engines. *Theory, Culture & Society*, 31(6), 125–150.
Wilson, K., & Keelan, J. (2013). Social Media and the Empowering of Opponents of Medical Technologies: The Case of Anti-Vaccinationism. *Journal of Medical Internet Research*, 15(5), 103.
Zimet, G. D., Rosberger, Z., Fisher, W. A., Perez, S., & Stupiansky, N. W. (2013). Beliefs, Behaviors and HPV Vaccine: Correcting the Myths and the Misinformation. *Preventive Medicine*, 57(5), 414–418.

APPENDIX: List of interviewees

Mittland County Council (the app)

Hanna, communicator (June 26, 2013)

Johan, health care planner (January 9, 2013)

Karin, gynecologist (February 28, 2013)

Katarina, information secretary (March 14, 2013)

Roger, health care planner (January 23, 2014)

Sara, school nurse (May 13, 2013)

Stefan, doctor, coordinator for the Infectious Disease unit (March 14, 2013)

Bredland County Council (the "I love me" campaigns)

Emma, epidemiologist (January 8, 2015)

Helena, communicator (February 18, May 24 and December 17, 2013)

Klara, Head of Communications, communicator (January 20, 2015)

Linnea, administrator for care choice system, nurse (February 18, 2013)

Previously published in the Pandora Series

Boel Berner, *Perpetuum mobile? Teknikens utmaningar och historiens gång* (1999)

Johan M. Sanne, *Creating Safety in Air Traffic Control* (1999)

Sabrina Thelander, *Tillbaka till livet. Att skapa säkerhet i hjärtintensivvården* (2001)

Kerstin Sandell, *Att (åter)skapa "det normala". Bröstoperationer och brännskador i plastikkirurgisk praktik* (2001)

Viveka Adelswärd & Lisbeth Sachs, *Framtida skuggor. Samtal om risk, prevention och den genetiska familjen* (2002)

Boel Berner (ed.), *Vem tillhör tekniken? Kunskap och kön i teknikens värld* (2003)

Christer Eldh, *Den riskfyllda gemenskapen. Att hantera säkerhet på ett passagerarfartyg* (2004)

Ericka Johnson, *Situating Simulators. The Integration of Simulations in Medical Practice* (2004)*

Tora Holmberg, *Vetenskap på gränsen* (2005)

Staffan Wennerholm, *Framtidsskaparna. Vetenskapens ungdomskultur vid svenska läroverk 1930–1970* (2005)

Elin Bommenel, *Sockerförsöket. Kariesexperimenten 1943–1960 på Vipeholms sjukhus för sinnesslöa* (2006)

Robert Hrelja, *I hettan från ångpannan. Vetenskap och politik i konflikter om tekniska anläggningar* (2006)

Jonas Anshelm, *Bergsäkert eller våghalsigt? Frågan om kärnavfallets hantering i det offentliga samtalet i Sverige 1950–2002* (2006)

Christer Nordlund, *Hormoner för livet. Endokrinologin, läkemedelsindustrin och drömmen om ett botemedel mot sterilitet 1930–1970* (2008)

Mats Brusman, Tora Friberg & Jane Summerton (eds.), *Resande, planering, makt* (2008)

Johan Wänström, *Samråd om Ostlänken. Raka spåret mot en bättre demokrati?* (2009)

Ingemar Bohlin & Morten Sager (eds.), *Evidensens många ansikten. Evidensbaserad praktik i praktiken* (2011)

Maria Björkman, *Den anfrätta stammen. Nils von Hofsten, eugeniken och steriliseringarna 1909–1963* (2011)

Annemarie Mol, *Omsorgens logik. Aktiva patienter och valfrihetens gränser* (2011)

Boel Berner, *Blodflöden. Blodgivning och blodtransfusion i det svenska samhället* (2012)

Olof Hallonsten (ed.), *In Pursuit of a Promise. Perspectives on the Political Process to Establish the European Spallation Source (ESS) in Lund, Sweden* (2012)*

Boel Berner & Isabelle Dussauge (eds.) *Kön, kropp, materialitet. Perspektiv från fransk genusforskning* (2014)

---

* Also available in an Arkiv Academic Press international edition, please visit www.arkivacademicpress.com for up-to-date information on available titles.

Previously published in the [illegible] series

[illegible]

Arkiv Academic Press

Arkiv Academic Press is an imprint of the Swedish publishing house Arkiv förlag. For up-to-date information on distribution and available titles, please visit:

www.arkivacademicpress.com

Published books

Ericka Johnson, *Situating Simulators. The Integration of Simulations in Medical Practice* (paperback 2012 [original edition by Arkiv förlag 2004])

Olof Hallonsten (ed.), *In Pursuit of a Promise. Perspectives on the Political Process to Establish the European Spallation Source (ESS) in Lund, Sweden* (paperback 2012)

Rebecca Selberg, *Femininity at Work. Gender, Labour, and Changing Relations of Power in a Swedish Hospital* (paperback 2012)

Sven E O Hort (birth name Olsson), *Social Policy, Welfare State, and Civil Society in Sweden*. Volume I: *History, Policies, and Institutions 1884–1988* (hardcover & paperback 2014, 3rd enlarged edition [1st edition by Arkiv förlag 1990])

Sven E O Hort (birth name Olsson), *Social Policy, Welfare State, and Civil Society in Sweden*. Volume II: *The Lost World of Social Democracy 1988–2015* (hardcover & paperback 2014, 3rd enlarged edition [1st edition by Arkiv förlag 1990])

Lisa Lindén, *Communicating Care. The Contradictions of HPV Vaccination Campaigns* (paperback 2016)

www.ingramcontent.com/pod-product-compliance
Lightning Source LLC
Chambersburg PA
CBHW020343270326
41926CB00007B/301